Noise

Noise

A Human History of Sound and Listening

DAVID HENDY

An *Imprint of* HarperCollins*Publishers*

HarperCollins books may be purchased for educational,
business, or sales promotional use. For information please e-mail
the Special Markets Department at SPsales@harpercollins.com.

Based on the BBC Radio 4 series.
The radio series was produced by Rockethouse Productions.

First published in Great Britain in 2013 by Profile Books LTD.

A hardcover edition of this book was published in 2013 by Ecco, an
imprint of HarperCollins Publishers.

FIRST ECCO PAPERBACK EDITION PUBLISHED 2014.

Library of Congress Cataloging-in-Publication Data
has been applied for.

ISBN 978-0-06-228308-5

14 15 16 17 18 OV/RRD 10 9 8 7 6 5 4 3 2 1

For Henrietta, Eloise and Morgan

Contents

Introduction

We're supposed to hate cacophony, but a few years ago on a cold Sunday in Berlin I was struck by the horror that sometimes lurks in silence and by the warm humanity that often emanates from noise. My teenage daughter and I had taken a suburban train north from the city centre to visit the old Sachsenhausen concentration camp in Oranienburg, where more than 200,000 people had been imprisoned under the Nazis. It was still early when we arrived, and only a few other people were around; a chilling mist, which clung to the place all morning, only added to the bleak atmosphere. Its utter noiselessness seemed oppressive yet entirely appropriate: whatever life the camp had once contained had been expunged cruelly many years ago. As the two of us walked around, looking at the evidence of one atrocity after another, it was difficult to know what to say to each other. So, like everyone else, we stayed silent.

After a few hours of this, and knowing we had to catch a flight home later the same day, we decided that we needed to cheer ourselves up pretty quickly. We caught the next S1 train back to the city centre and made our way to Café Einstein for cakes and coffee. The moment we stepped inside this venerable wood-panelled Weimar institution, crammed to bursting point with Berliners having their Sunday afternoon treat, we were hit by an extraordinary wall of sound. The idea of finding somewhere quieter never occurred to us, however. The clatter and clinking of cutlery and crockery as waiters hurried from table to table, the ringing of tills, the shouting of orders from the kitchen and, rising above everything else, the constant loud buzz of conversation and laughter coming from everyone: after a long morning's silence this din was a blissful affirmation of life, a sonic two-fingers to the Nazis and the deathly silence they had created at Sachsenhausen.

Noise, it has been said, is sound that is 'out of place'.[1] It is usually something unwanted, inappropriate, interfering, distracting, irritating. Many of us would no doubt concur with the nineteenth-century German scientist Hermann von Helmholtz, who distinguished clearly between 'musical tones' and mere 'noise', the latter being sounds that are all 'mixed up and as it were tumbled about in confusion'.[2] But on that day in Berlin I saw, as I hope to argue in this book, noise is more important than this. When the bell rings, a factory siren sounds, or the skies fall silent after a terrorist attack, noise – or its absence – is charged with meaning. Noise has been a capacious category throughout human history – one full of surprises and drama.

I am with John Cage. 'Wherever we are what we hear is mostly noise,' he wrote in 1937. 'When we ignore it, it disturbs us. When we listen to it, we find it fascinating.'[3] If we open our ears to sounds that are usually dismissed as unmusical or unpleasant, or simply ignored as merely everyday and banal, Cage implies, we start reconnecting with a whole range of human experience that previously passed us by. Instead of worrying about the usual boundaries between noise and music, or cacophony and silence, or speech and song, we need to discover the virtues of breaking them down.[4]

So although this book has the word 'noise' prominently in its title, it is trying to stretch the definition as far as it will go – and in lots of directions, too. It encompasses not just music and speech but also echoes, chanting, drumbeats, bells, thunder, gunfire, the noise of crowds, the rumbles of the human body, laughter, silence, eavesdropping, mechanical sounds, noisy neighbours, musical recordings, radio, in fact pretty well anything that makes up the broader world of sound and of listening. When I turn to oratory in ancient Rome and in modern political campaigning, for example, I am interested in the words spoken but I am even *more* interested in the sounds made: the tone, the cadences, the pitch of the voice; how that voice might have been transformed by the environment in which it was heard, and how the audience responded. When I discuss the jazz scene in Harlem during the 1920s, it is less the musical quality of Mamie Smith or Ma Rainey that concerns me and more the impact recording had in allowing 'new' sounds to circulate well beyond a small group of people gathered at a concert or dance-hall and in allowing the 'voice' of a marginalised culture to be 'heard' as never before by an international listenership.

Having said all this, I still want to hold on just a little bit to that original idea of noise as a nasty, troubling thing. For although I think noise is not always a sound 'out of place', nor always strictly speaking unwanted, it can perhaps be thought of as a sound that someone somewhere doesn't want to be heard. By that, I mean that who gets to make a noise and who doesn't, who gets their voice heard and who doesn't, who gets to listen and who doesn't, is of crucial importance. Silence can be golden, or it can be oppressive. And as the history of slavery, or the history of the relationship between factory-owners and their workers shows us, whether it is enforced or voluntary makes a world of difference. So this book is really about how sound might help us understand some of the drama and struggle of human history in a new and, I hope, enlightening way.

To trace the story of sound is to tell the story of how we learned to overcome our fears about the natural world, perhaps even to control it; how we learned to communicate with, understand and live alongside our fellow beings; how we have fought with each other for dominance; how we have sought to find privacy in an increasingly busy world; how we have struggled with our emotions and our sanity. It encompasses the roar of the baying crowd in ancient Rome, medieval power struggles between rich and poor, the stresses of industrialisation, the shock of war, the rise of cities, the unceasing chatter of twenty-four-hour media. Throughout all this, we have to keep our ears attuned to the intimate aspects of human life as much as the epic. For, as the historian Elizabeth Foyster reminds us, senses such as listening have always been a part of our private domestic life, our thoughts, our feelings, our memories; in other words, 'a crucial part of the everyday'.[5]

I keep using the word 'human' for a reason. It is to mark out a subtle but important difference between this book and most other work written on sound. Hillel Schwarz's *Making Noise: From Babel to the Big Bang and Beyond*, Veit Erlmann's *Reason and Resonance: A History of Modern Aurality* and Mike Goldsmith's *Discord: The Story of Noise* are just three among several recent contributions to the new frontier of 'sensory history', all of them deeply fascinating.[6] But they are written from the perspectives of, respectively, a poet, a music anthropologist and a physicist. Though they discuss people – how could they not when dealing with sound? – their main focus, it seems to me, has been with sound as an idea or a metaphysical phenomenon; they offer what is essentially an intellectual history of the subject. Valuable though that is, my own interest lies less with the abstract or physical qualities of sound than with how it gets used in the world by you and me and everyone else. In other words, I am interested in its *social history*, and, equally important, in the history of how and why we have listened to it and reacted to it.

This means a special fascination in what follows with the subjective aspects of sound: what it actually *felt* like to experience certain sounds in certain places at certain times in history. The pioneers in this respect have been historians such as Alain Corbin in France, and Mark M. Smith, Richard Rath and Emily Thompson in America. Their approach, as Thompson puts it, has been to proceed on the basis that, like a landscape, 'a soundscape is simultaneously a physical environment and a way of perceiving that environment'. 'It is,' she suggests, 'both a world and a culture constructed to make sense of that world.'[7] These historians take the pioneering idea of the Canadian musician R. Murray Schafer, who first popularised the term 'soundscape' in the 1970s, and test what exactly that meant for ordinary people in very particular times and places: Corbin explored the role of church bells in nineteenth-century rural France; for Smith it was the sounds of the slave plantations and battlefields in nineteenth-century America; for Rath, the drums and guns of colonial-era America; and for Thompson, the cityscapes of the early twentieth century. This work, and other work like it, provides some of the essential building blocks of the story presented here.

But I want to offer, if I can, a wider story, both chronologically and geographically, for, as Richard Rath suggests, a simple noise such as thunder will have been interpreted very differently by, say, Native Americans and New England colonists. I would add that, most likely, it would have been heard very differently by the early humans of the Palaeolithic, by ancient Greeks, by medieval monks or by soldiers in the First World War trenches of Flanders too, though I should hasten to add that it would sometimes have been heard in very *similar* ways, since we find, for example, that medieval monks and nineteenth-century French farmers – both equally irrationally – viewed thunder as having a supernatural force behind it. Which is to say that one of the benefits of pursuing a history that stretches all the way from prehistory to the present, and encompasses several different parts of the world, is that, whatever is lost in terms of detail, we can at least start to tease out a few continuities, as well as identify a few dramatic breaches, in the long story of our relationship with sound.

This is important because the history of the relationship between sound and human history has tended to be told almost entirely in terms of a quiet 'then' and a noisy 'now'. When exactly 'then' *is*, is of course debatable – as is the perceived cause of any rupture. The most common account puts the Industrial Revolution centre stage. This was the position of the Glaswegian doctor Dan MacKenzie, the writer in 1916 of the allegorical *City of Din*. 'Nature,' the doctor argued, was 'quiet' and 'pleasant'; modern civilisation, on the other hand, 'is noise. And the more it progresses the noisier it becomes.'[8] This, broadly, was also the line taken by R. Murray Schafer in the 1970s, when he declared that the sounds of nature had been 'lost under the combined jamming of industrial and domestic machinery'.[9] It's a line that pits the natural world and humanity against each other, and it retains a strong appeal to environmentalists. Yet I worry about it edging into slightly misanthropic territory, as if the world would be better if only the people in it disappeared. And, as Emily Thompson has suggested, there is an equally strong case to be made that soundscapes have 'more to do with civilization than with nature'; indeed, that our soundscapes are constantly changing in subtle ways, and not always for the worst.[10] This, I hope to argue, is emphatically not a simple story of irreversible decline into ever greater cacophony.

A rather different way of dividing the human timeline has been to distinguish between an 'oral' then, which was somehow more magical than the present, and a 'literate' now, which is somehow more rational than the past. In effect, this divides 'ear' culture' (listening) from 'eye' culture (watching and reading) and then proceeds to show that once reading had taken over, 'the visual' came to be regarded as the more comprehensive and trustworthy sense, while 'the aural' was left behind, with associations of passivity, superstition and hearsay. According to taste, this fundamental shift happened either in ancient Greece, when writing was systematically adopted, or during the Enlightenment, when the habit of reading spread rapidly. Even if we take this theory at face value, it's worth pointing out that a truly global, multicultural perspective, which anthropologists are good at providing for us, shows that a 'pre-literate' society is something that continues to exist long into the 'modern' era. But why take it at face value? We surely need to question almost every assumption that has been made here about the supposed triumph of a visual sensibility as time passes, and about the consequent relegation of aural culture: that hearing is less important now than it has been in the past, that listening is a passive activity, that seeing something provides better proof than hearing something, that what happened in the West also happened in the East. A social history of sound and listening suggests otherwise.

But suggests what, exactly? I hope that the following chapters can, to some extent, simply be allowed to tell a series of separate stories. Even with thirty chapters, the span of humanity covered is too great to offer more than a few snapshots, and sound, especially, is too profuse a subject to pin down into a single, coherent narrative. Yet I suppose there *is* a running thread of sorts: it is about power. I mean this in two senses. First, it's about the power of certain sounds to influence us in profound ways. And secondly, it's about the ability of powerful people – or powerful groups of people, like nation states, organised religions or commercial companies – to shape the soundscapes or listening habits of others less powerful. One of R. Murray Schafer's great contributions to our understanding of the subject was to think of sound as a way of *touching at a distance*. His notion captures perfectly the way that sound travels further than the length of an arm but arrives in someone's ear as a tangible thing, triggering a real emotional response. It is therefore a force acting upon people, for good or ill. At the same time, sound never bestows absolute power on anyone, since by its very nature it is hard for sound to be entirely owned or controlled. Its natural tendency is to move freely through the air. And although human ingenuity is such that sound can always be manipulated, sound is also too intangible and slippery a thing to remain in the service of elites without also being available for use in inventive and subversive ways by the dispossessed – as a brief history of medieval carnivals, eighteenth-century rebellions and twentieth-century protest marches will show.

Being intangible and slippery, one might easily imagine that sound is almost impossible to write about in the purely historical sense. As Douglas Kahn points out, 'Sound inhabits its own time and dissipates quickly.'[11] It leaves no traces, and the discipline of history needs traces. That is why historians spend their time among written archives: they provide a satisfyingly stable record of what happened in the past. Yet it turns out that many sounds, even of the distant past, are not entirely lost to us. We can make some sensible guesses about them if we deploy a bit of sideways thinking. Archaeologists, for instance, have begun using experimental techniques to explore the acoustic properties of ancient sites. They also now draw on ethnographic studies of present-day hunter-gatherer societies in order to speculate on the possible human uses of sound in prehistory. In doing so, they have invented a whole new discipline, 'archaeoacoustics'. Historians of later periods have also turned to anthropology and ethnography to help them understand past behaviours, such as the role of eavesdropping in different cultures or the effects of overcrowding. Indeed, it is the fieldwork of ethnographers that has helped, more than anything, to build up today's voluminous archives of sound recordings, such as the British Library's collection of several million wax cylinders, discs, tapes and CDs, which now allow us to bring back to life an array of voices and music and soundscapes from over a hundred years ago.

Finally, though, we should not forget that even our most traditional source for history, the written record, sometimes tells us a great deal about the sounds of the past. In countless letters, journals, diaries, speeches and books, people from every period of history and every part of the world have recorded their personal impressions of places and events. In doing so, they frequently wrote not just of what they saw but also of what they heard. Sometimes this was because what they heard struck them as extraordinary and so deserved to be recorded in detail; at other times, the references are incidental and fleeting – but, for us, no less informative. That so many people chose to write about sound is a clear measure of how important it was in their lives. And what these people tell us, in the pages that follow, is this: that the desire to understand and control sound – to enforce silence, to encourage listening, to sing, to shout – is not just hundreds but tens of thousands of years old.

I

Prehistoric Voiceprints

1

Echoes in the Dark

If you have ever been into one of those preserved caves that our prehistoric ancestors visited, you will know that two things usually happen at once. You are pretty quickly smothered in complete darkness, and you suddenly leave behind the sound of the outside world. A blissful respite from the noise and bustle of modern life, you might think. In fact, it's far from silent and peaceful. As a listening experience, it can even be quite unnerving.

During the Middle and Upper Palaeolithic, some 40,000 to 20,000 years ago, small groups of men, women and children – Neanderthals at first, then our most direct ancestors – would have gathered near the entrances of caves across Western and central Europe for shelter, and perhaps gone deep inside for rituals. These enclosed spaces have their own acoustic character:

echoing voices, of course, but also intensifying them. If you visit them today, you will notice that every sound you make as you walk through them lingers longer, reverberating, and coming back to you from unpredictable directions, thanks to the irregular shape of the walls.[1] In certain places there is a cacophony of echoes – each one lasting long enough to merge with the next to create an almost continuous wall of sound, rich, complex and, to the untrained ear, pretty disorientating. When we whisper, hum, speak or sing, they shout and sing back to us. These caves are alive.

Perhaps it's not all that surprising that caves resonate. But several archaeologists have tried an experiment that reveals something rather more remarkable. Moving slowly, and in total darkness, along the narrower passages of caves such as Arcy-sur-Cure in Burgundy, and Le Portel near the Pyrenees, they have used their voices as a kind of sonar, sending out a pulse of sound then listening out for any unusually resonant response. Most of us can do this, by the way: almost without noticing it, we tend to use subtle cues such as variations in loudness and variations in the time of arrival at our ears of different echoes to very swiftly 'localise' sound – to navigate, in fact, a bit like bats in the night sky.[2] The point, in any case, is that when these archaeologists felt the sound around them suddenly changing, they would turn on their torches. And at that precise point they would often see on the wall or ceiling a painting. This might be something as simple as a small dot of red ochre. Or it could be more complex – a pattern of lines, a negative handprint, an animal.[3] What is significant is that wherever a cave *sounds* most interesting, you are also likely to find the greatest concentration of prehistoric art.

The first person to map in detail this stunning coincidence of resonance and art was the musicologist Iégor Reznikoff. After walking carefully through the caverns and tunnels of Arcy-sur-Cure for himself in the mid-1980s, and making a detailed map of what he heard and saw, he reckoned that about 80 per cent of the images are in spots where the acoustics are particularly unusual.[4] For example, near the bottom of a cave called the Grand Grotte, where each sound might provoke up to seven echoes, there are paintings of several mammoths, some bears, a rhinoceros or two, a salmon, some sort of cat and an ibex. And in a mezzanine area near the so-called Salle des Vagues (the 'Hall of Waves'), just where the resonance is really striking, there's a ceiling densely packed with animals of all kinds, and, on the floor, the delicate outline of a bird. At other caves there's the same pattern: at the cave of Niaux in the Pyrenees, for instance, almost all the animal paintings are in the Salon Noir, which Reznikoff describes as sounding like a richly resonant Romanesque chapel;[5] and at Le Portel, a whole series of red dots runs along a ten-metre tunnel, each one, again, precisely where, as Reznikoff puts it, a 'living sound point lies'.[6]

Why didn't the artists who made these prehistoric paintings work nearer the cave entrance, where there's much more space and light? We don't know for sure: it's impossible to guess their thoughts. But clearly *something* drew them to the darkest, deepest and most inaccessible parts of each underground complex. Even prehistoric art that has been found outside caves is sometimes located in inconvenient places: high on canyon walls and cliff faces. Again, it's crowded on to some surfaces while other rocks nearby are left strangely blank. And again, it's sound that seems to provide the link.

Go rock-art hunting in Horseshoe Canyon, Utah, or in Hieroglyph Canyon, Arizona, for instance, and you'll find that those places with the greatest concentration of pictures – human figures, mountain sheep or deer – are exactly the same places where echoes are strongest or where sounds carry furthest.[7] The connection between the sound quality of a particular spot and the art that is nearby just keeps cropping up.

So much so that it's a good guess that our prehistoric artists didn't select by accident those surfaces – whether deep inside a cave or high up on a cliff – that created the most interesting acoustics. They seem to have chosen them deliberately – as if they couldn't shake these echoes out of their minds.

What, then, was going on? Why did the sound of an amplified echo apparently fascinate prehistoric peoples so much? One clue has emerged at the Music School in Cambridge, where an intriguing experiment was conducted in 2000. The musicologist Ian Cross, the anthropologist Ezra Zubrow and the archaeologist Frank Cowan came together in an open-air courtyard to practise the prehistoric craft of flint-knapping. Bone pipes or flutes excavated from various sites in Europe had already shown that humans were making music from about 36,000 years ago. But what about before then? The three investigators wondered if even older, stone objects might also have been used to make music.[8] They tried holding the flints and striking them in different ways, and they soon discovered an array of sounds could indeed be made.

It was impossible to prove that these sounds were actually exploited by prehistoric peoples for anything we might recognise as 'music'. But in the middle of all the testing something unexpected happened. A stone blade being held between two fingers was tapped, and the three men in the courtyard suddenly heard a high-pitched flutter – what sounded very like a bird nearby flying away from them. Though they were out of doors and in the full afternoon sun, Ian Cross recalled the effect as being 'quite unearthly … it seemed that the tapping had suddenly awoken some real yet invisible entity' – like an avian spirit.[9] He knew there was a perfectly good scientific explanation to hand: the shape of the courtyard, the mix of building materials, the sound produced, the men's position – all this had set up a pattern of sound waves, which created a moving, fluttering echo with a life of its own. For the rest of the afternoon they tapped the stone blade again and again, and discovered that, given the right mix of circumstances, they could keep evoking the sound of a bird flying across the courtyard. They knew there was hard science behind the phenomenon. But they claimed this 'did nothing to dispel the "magical" qualities' of the fluttering sound they'd created.[10]

What is most interesting about the Cambridge experiment isn't just the creation of a special effect. It's the idea of an invisible animal spirit having been unleashed through sound. In fact, at many prehistoric sites, echoes conjure up something similar: when a clap in a cave bounces back in a series of overlapping echoes, it's not so much the cave that comes alive, but the animals painted or engraved on the walls nearby. They gallop and stampede about, as if the sound of hooves really were coming from within the walls themselves. The sound isn't just sharing its space with the image; it's *mimicking* it. Or perhaps the image is mimicking the sound. At other times, a noise made in one place appears to be answered from somewhere else entirely. Occasionally, a sound might seem to come from behind a rock rather than from its surface, as if its point of origin were deep within or the surface itself were a chimera. All these effects are uncanny. Prehistoric people would have had no understanding of the science of sound waves and reverberation. For them any echo would surely have seemed like a new sound, coming from some invisible being or spirit – something, perhaps, from within the rock, speaking back, making its own presence felt.[11] In other words, it would have seemed supernatural.

And sure enough, if we look at different cultures around the world, time and time again we find myths involving supernatural echoes – myths with their roots almost certainly deep in prehistory. Among the Native American Paiute people, for example, there are stories of witches living among the rocks, taking great delight in repeating the words of passers-by.

Among the Cherokee, there are countless names for rocks that 'talk'. In southern Africa, the San Bush people, who have been producing some of the world's greatest rock art for thousands of years, often use images showing figures and patterns crawling out of the cracks or holes of the stone, as if emerging from a teeming spirit world just 'behind' the surface.[12] It's hard to resist this thought: that places which echoed were special – 'labelled' by these painted images as being full of spirits, as sacred places.

There's also an intriguing connection with music, and, through music, to trance. In San rock art, one recurring image is of human figures dancing in a kind of trance state; others include monsters, fish, eels, turtles and the eland. The archaeologist David Lewis-Williams believes these paintings might represent the visions of those in a trance – what they witnessed when they lifted the '"veil" suspended between this world and the next'. The images also make sense *because* they're so often found on the walls of rock shelters: these resonant surfaces – walls that seem, from the sound they make, to be inhabited – are, in effect, the very gateways to this spiritual realm.[13]

So perhaps prehistoric people, when they went into caves like the ones at Arcy-sur-Cure, Le Portel and Niaux, weren't just there to stand passively, transfixed in wonder at the strange sounds stirred up by their presence. They might have been going in to actively *invoke* a spirit world, to be in dialogue with it through creating their own noise and listening to the results: tapping flints, perhaps, to set off fluttering echoes, or even hitting a pillar of rock.

We can hear musical stones being played across the world: the 'pichanchalassi' lithophone (or musical stone) in Togo, 'gong rocks' in Namibia, and 'ringing rocks' in southern India, Scandinavia and North America.[14] In one way or another, the sound of the lithophone is ubiquitous – and probably has been for most of human history. So it's perfectly possible that these ringing sounds could also have been drifting through European caves tens of thousands of years ago.

Certainly, in caves at Roucadour, Cougnac and Pech-Merle in France, at Nerja in Spain and at Escoural in Portugal, there are rock pillars decorated with red dots and bearing all the marks of being repeatedly hit. Some of them even give off differently pitched sounds when struck.[15] These are also the caves that have left behind some of the world's oldest surviving musical instruments, the bone pipes or flutes mentioned earlier.[16] If found in caves, they were probably played in caves. Indeed, some of the very oldest bone flutes, discovered at Isturitz in the Pyrenees, were found next to a decorated pillar, in the one chamber that amplified sound more than any other part of the cave. They are yet more evidence that some kind of music was an important element in what humans did in such places roughly 20,000 years ago during the Upper Palaeolithic. But prehistoric people didn't really *need* to 'invent' bone pipes in order to make music in here. They already had stone pillars to hit, their own voices and, of course, the wonderful resonance of the caves or rock shelters themselves. Here, in the dark, with only the flickering half-light of their lamps or tapers, the atmosphere would surely have been perfect for rituals or celebrations, for music and singing, for summoning the supernatural.

In the midst of such apparent magic, our ancestors must have wanted to keep making sound, if only to keep the conversation with the spirit world going too. So we can begin to see that through noise we evolved. In a continuous feed-forward loop, new sounds, tonal effects, notes and rhythms were discovered. They were tried out, they echoed back, they were copied, altered, replayed, thousands of times, over and over again. And, eventually, from chaos emerged order.[17]

Of course, all this chanting and playing wasn't just about

communicating with a spirit world. Often, it was about communication between living people in *this* world – about men and women and children doing something together in time, about bonding, sharing. Which is why, to help us understand the distant origins of both language and family life, we have to turn to the beat of African drums.

2

The Beat of Drums

One of the great treasures of the British Library's sound archive is a scratchy wax-cylinder recording, made in 1921 by Captain Robert Sutherland Rattray, a British colonial administrator. He lived among the Ashanti people of Ghana, and wanted to capture a particularly remarkable aspect of their lifestyle. His recording is one of the very earliest made of the 'talking drums' of Africa – drums made out of tree trunks and struck with two wooden sticks, one in each hand. The drum itself is hollowed out, so that its shell is left thicker on one side than the other. Just as Morse code is made up of dots and dashes, the talking drum gives out a high tone or a low tone, depending on where exactly it's being hit. A precise combination of these different tones makes up the message, which then travels like a Morse-code signal pulsing along an invisible telegraph wire through the dank, dark equatorial rainforest.

This is a place where it's impossible to see very far: the only way to communicate is by sound. And the talking drum hurls its powerful and complex rhythms into the air along a radius of perhaps six or seven miles – much further, of course, if each message is relayed not just once but is repeated, from village to village, rumbling through the trees, over hills and along rivers, all the time faster than anyone could run. As James Gleick points out, for hundreds if not thousands of years, 'no one in the world could communicate as much, as fast, as far as unlettered Africans with their drums'.[1] But this isn't just a striking example of clever communication by sound, a form of wireless telegraphy before the invention of wireless telegraphy. When asked how long they have been using the talking drums, the people of West Africa say, 'We have always had the drum.' It's a tradition that not only tells us about communication, but also helps us travel back in time to reveal the vital role of sound – and, in particular, of rhythm – in human evolution.

Captain Rattray, who made that recording in wax, wasn't the first westerner to notice the talking drums. In the seventeenth and eighteenth century, slave traders and Christian missionaries had heard their insistent rhythms, too. They interpreted them, rather nervously, as a call to fight, or maybe as some 'Hellish' pagan custom or immoral merrymaking. In any case, they didn't want to enquire much further.[2] By the 1920s and 1930s, however, Rattray was part of a new generation of more enlightened settlers trying to learn a little more about the people among whom they lived. Another member of this invading western tribe was a missionary called Roger Clarke, based among the Tumba tribes further inland, near the Congo River. Like Rattray, Clarke was no longer content just to marvel at the drum language or to live in fear of it; he wanted to crack wide open its secret code.

After listening attentively with the help of local translators, he concluded that the drum language consisted mostly of curiously long-winded messages. For example, a simple call to fight would, through the talking drums, become something like this:

> Make the drum strong; strengthen your legs, spear, shaft and head, and the noise of moving feet; think not to run away.[3]

And a farewell to the sun in the evening would become this:

> Shining sun, who has made a dwelling in the sky, who has gone to the concourse of counsel, all morning, all day, evening comes, you are going, good-bye.[4]

This is all richly poetic. But Clarke was somewhat mystified by the drum's sounding out of each message taking so much longer than if the message had just been spoken or shouted out loud. Why this inefficiency?

The answer came a few years later from another missionary who had been eavesdropping in the rainforest, John Carrington. Part of the problem, Carrington realised, was that when everything was reduced to two-tone beats, there was lots of potential for confusion. In the Kele tribe version of drum language, for instance, a double stroke on the high-tone lip of the drum could mean moon, bird fowl, a kind of fish or countless other things. Add extra drumbeats or phrases though – 'the moon looks down at the earth', 'the fowl, the little one that says kiokio' – and ambiguity evaporates. Carrington and his wife had learned the technique for themselves. So when she wanted to summon him from the forest for lunch she would beat out the following message:

> White man spirit in forest come to house of shingles high up above of white man spirit in forest. Woman with yams awaits. Come come.[5]

This was functional, of course – but also, one hopes, in its delicious verbosity, an expression of love.

The basic mistake of the earlier settlers had been to think of the drum language as simply a way of trying to 'get across' a message from a sender to a receiver. Actually, it's never been a form of signalling; it's really a whole language, and it's used in a conversational way – chatty, informal, jokey, interactive, taking turns. And in conversation, the process of talking to each other – and the manner in which we do it – is almost always as important as what is being talked about. When it works, all that back-and-forth rhythm, with the subtle shifts in tone, weaves a special kind of magic, getting us 'in tune' with each other, pulling us closer together.

There's nothing peculiarly African about any of this. It's a technique that we can find across a wide range of human behaviours. 'Baby talk', or what linguists call 'infant-directed speech' – that sing-song voice care-givers use, without thinking, when chatting to their babies – is an example of where the melodic and rhythmical nature of speech, its sheer emotional expressiveness, clearly counts for much more than what is being said. (Babies, of course, can't *really* understand any of the words, and certainly can't yet talk back.) If we were to eavesdrop on midwives, for example, anywhere in the world, we would hear pretty much the same thing. The melodies and intonations of 'infant-directed-speech' are virtually identical: it's a universal phenomenon.[6] So, even though this kind of interaction might help babies learn language, that is not the main point of it. The melody is the message: what is being transmitted is *emotion* – and, through that, the building up of a strong bond, in this case between adult and child.

Sounds provide a means of us 'touching' at a distance – a form of personal contact that can work even when we are physically beyond the reach of one another. So even a dialogue between teenagers texting one another on their mobile phones is all about creating a kind of social glue – social glue that works through the hidden melody and rhythm of that constant toing and froing of words. Such musical qualities are deeply embedded in the way we talk. So deeply embedded, so universal, indeed, that they give a strong hint about the past: that if we go back, say, to around two million years ago, to a point before we had language or even music – before culture had emerged and started to create an infinite variety of expressive forms in human life – we might find that the very first proto-humans to leave Africa had something else instead. This was something with elements of both language and music, but which was not quite either. It was a kind of sing-song utterance that has been called 'Musilanguage'. The possible existence of this strange noise is why one leading archaeologist, Steven Mithen, refers to some of our distant ancestors as the 'Singing Neanderthals'.[7]

One way of getting a sense of how this might have sounded would be to turn to creatures even more distantly related to us: African apes and monkeys. Vervet monkeys, for instance, have a thrilling range of alarm calls – calls that are apparently communicating quite precise information about the kind of predator or threat that has been spotted nearby. The calls of gelada monkeys, from the Ethiopian highlands, are less distinctive, but what their communication lacks in data, it more than makes up for with melody, which is quite complex and rhythmic. Gibbons are less chatty, but no less musical in their own way – males and females sometimes performing a kind of duet. The grunts and hoots of gorillas, chimps and bonobos sound a bit restricted in comparison with all this. But then they use physical gestures as well as vocalisations to get their message across. Their overall toolkit of techniques is bigger. It's with these apes and monkeys that we shared a common ancestor several million years ago, so it's reasonable to guess that the shrieks are very roughly the kinds of sounds our own human ancestors would also have been making.

When they needed to work together as a group – during, say, a hunt – the ability of these early humans to coordinate was vital. In the dark world of the rainforest this could only be done through sound. We can be pretty certain about this because of the compelling evidence provided by the behaviour of modern humans in the equatorial forests of Central Africa. The BayAka pygmies of the rainforests of the Central African Republic, for example, perform boyobi ceremonies, just before setting off to hunt. The men chant and drum, while choirs of BayAka women weave shimmering webs of polyphonic sound to entice the bobé forest spirits to dance and bless the hunt with symbolic spit. Wearing leaves, the bobé appear and dance. But if the bobé spirits don't think the musical skills on display are good enough, they will harangue the hunters in falsetto voices, bringing the group's rhythmical efforts to a crashing halt. Sing better, drum better, and you will eat tomorrow! There's lots of information in these sounds. But what is really being created is *coordination*: listening, calling and responding, taking turns.

It's these men's ability to fall into a shared rhythm that is vital to the success of their hunt; it's what tests their fitness to act together in all sorts of ways. By making noise and moving together in time, the group isn't just working well on the occasion of this hunt. Each time it performs the ritual, it's blurring self-awareness, building up trust, creating a stronger social bond based on shared emotions. The term usually given to this phenomenon is 'entrainment'.[8] It's what happens when different rhythms start to interact and then to synchronise. It's listening to music and starting to tap our feet or fingers in time. Or the thumping soundtrack of an action film quickening our pulse as we sit mesmerised in the cinema. It's a group of soldiers adjusting their walking pace to march in unison. Or the repetitive movements of workers on a factory production line, lock-step in time with the robotic machinery around them. In each case, an individual bodily rhythm is being sub-sumed into a collective rhythm. This is why for those with Par-kinson's disease one way of compensating for a loss of control over movement is to be guided – regulated, even – by some external source of regular beats. It doesn't always have to be a rhythm imposed on us by others, though. Often, entrainment is more interactive and organic, with every participant con-stantly recalibrating to keep in unison. In such circumstances, it's impossible to tell the difference between performer and listener, simply because the distinction doesn't really exist. One reason why this falling into time often occurs so easily is that many of the coordinating rhythms are universal: the beat of a heart, the in-and-out of breathing, the steady gait of walking. It was almost certainly these simple biorhythms that continued to shape the undulating patterns of music and lan-guage, even when our distant ancestors had left the rainforests of Africa and moved into the open savannahs beyond.

In this new environment of the savannah, hunting, cutting meat and making stone axes would have been never-ending daily tasks. It's not hard to imagine the scene. Proto-humans in family-like groups, perhaps humming or cooing as, search-ing for food, they moved beyond each other's physical reach but still wanted to stay somehow connected. Or small groups crouching together on the ground, the insistent rhythms of their work occasionally drifting imperceptibly into singing, perhaps even dancing. And, as they sang and danced, and hummed, and chipped away, a vague sense of fellow feeling hanging in the air.

In situations like this, those who were most skilled in con-veying their own feelings to others, and in reading the feel-ings of others, would have been the most useful members of the community. Predicting the behaviour of others and being able to manipulate their behaviour was, in evolutionary terms, undoubtedly a big advantage. Something with the quality of music might well have been a safer evolutionary bet than using words. Words have always been very specific in their meanings. They usually spell out exactly how one feels and what one thinks. Those listening might agree with what is being said; equally, they might disagree and so quickly fall out with the speaker. Music is different. Its meaning is helpfully vague: we can interpret it to be almost anything we want it to be. Through its ability to 'entrain', music is also powerfully bonding. We can be swept along by it, forgetting ourselves a little. Which is why, even today, if an argument is brewing, friends and family members often find it's better not to say anything; they might instead find an excuse to put on some music, perhaps even sing along a bit, and wait until it all blows over.

For our distant ancestors, then, the strange musical vocalisations which archaeologists reckon probably came before 'proper' language quite possibly helped paper over any cracks that were appearing inside the group. These musical sounds would also have been a potent symbol of the group's togetherness in the face of any external enemies – an audible sign of its ability to 'engage in complex, collective and coordinated action'.[9] It was, in other words, a brilliant means of saying to outsiders, *Look what we can do: attack us if you dare.*

By the time early human groups such as this had reached all corners of Europe, it was still several hundred thousand years before humans were capable of sophisticated cave paintings, let alone the coded language of the talking drums. But these men, women and children were an early sign of what was to come. They give us tiny hints of the vital role that was going to be played by rhythm in driving forward our evolution.

Rhythm remains deeply rooted, a universal feature of human-made sound. It's why we still hear beguiling similarities in music from very different parts of the world – and why we still hear the same underlying melodies at work in completely different languages. But of course, no language or music is ever exactly the same. Over the past million years, as our predecessors spread further and further across the world, their cultures also inevitably drifted apart. In particular, the way they sang, made music or talked would become adapted to the different habitats in which they lived. Nowadays, we like to think of ourselves as having risen above nature. But as it turns out, the wilderness we imagine our ancestors left behind long ago is still deeply embedded in the sounds we make to this day.

3

The Singing Wilderness

In the northern woods of Minnesota, USA, lies one of the less-well-known wonders of the natural world: Burntside Lake. In the middle of this idyllic setting, facing out across the lake and surrounded by willow, birch and weathered pine, there's a small wooden cabin, built in 1956 by the American conservationist Sigurd Olson. In giving this private retreat a name, Olson avoided the predictable. He rejected 'Lake View' or 'Dunroamin', and called it, instead, 'Listening Point' – for this was, he wrote, a place where he could 'hear all that was worth listening for'. It was here, more than any other place, that he could experience a 'sense of oneness', simply because it was here more than any other place he knew of that there were 'no distracting sights or sounds'. Rocks, trees, water: these features of the landscape were all well and good. But for Olson the essence of any wild place was its silence. If 'Listening Point' was for listening to anything, it was to the very noticeable *absence* of noise.[1]

But was Olson right? Is silence really the essence of the world's wild places? It is true that here you will find an absence of *human* sounds. But nature itself can be immensely noisy – even in the wooded seclusion of a lakeside cabin. Evergreen forests give out what has been described as a 'low, breathy whistle'. If the wind gets up, they will seethe and roar and creak as branches rub together and millions of needles 'twist and turn in turbine motion'.[2] The deciduous woodlands of England have been described eloquently by Thomas Hardy: the holly whistling, the ash hissing, the beech tree rustling as its boughs rise and fall, and then, as winter comes and the leaves are shed, there's a subtle shift in notes.[3]

Keen-eared writers keep noticing this musical quality wherever they go. In 1874, the conservationist John Muir was deep in the Sierra Nevada forests. As the rain poured down, he heard Aeolian music emanating from the 'topmost needles'.[4] The tropical rainforests of the Amazon or Africa or Papua New Guinea are denser still, their storms more violent. When it rains in any of these places, an extraordinary cacophony of sound can be unleashed. The great collector of natural sounds, Bernie Krause, described being caught in an afternoon downpour as he set up his recording equipment, witnessing the first 'dense wall of water' falling like a 'freight train approaching', then, a few moments later, the melodic drips on the leaves, the odd splash of water hitting small pools on the ground, and finally, as the storm itself receded into the distance, insects beginning to stridulate – just a few at first, then in their thousands – and countless exotic birds starting to call out to one another, their cries reverberating as if in a giant cathedral.[5]

At moments like this the forest is wide awake. But really it's never fully asleep, because, of course, it's densely populated with living creatures. Which is why, when the young anthropologist Colin Turnbull first went into Congo's Ituri Forest in the 1950s, he discovered not the silence he had been expecting from having read the textbooks, but a rich sonic tapestry that was, in his words, 'exciting, mysterious, mournful, joyful':

> The shrill trumpeting of an elephant or the sickening cough of a leopard (or the hundred and one sounds that can be mistaken for it) ... At night, in the honey season, you hear a weird, long drawn-out, soulful cry high up in the trees. It seems to go on and on, and you wonder what kind of creature can cry for so long without taking breath ... Then in the early morning comes the pathetic cry of the pigeon, a plaintive cooing that slides from one note down to the next until it dies away in a soft, sad little moan. There are a multitude of sounds, but most of them are as joyful as the brightly coloured birds that chase each other through the trees, singing as they go; or the chatter of the handsome black-and-white Colobus monkeys as they leap from branch to branch ...[6]

No wonder the Earth has been described as a 'macrocosmic musical instrument', the creatures on its surface a 'great animal orchestra', pulsing with sound and rhythm.[7] Even Sigurd Olson, smoking his pipe in the relative quiet of Burntside Lake, Minnesota, wrote of a '*singing* wilderness'.

It's also clear this wilderness sings with a dazzling range of voices. It mutates from hour to hour and from season to season. Its register shifts from place to place. John Muir claimed on his hikes through the Sierra Nevada mountains that he always knew exactly where he was based solely on sound. Each square mile of forest has always had its own acoustic signature. As has each stretch of coastline, each bend in a river, each expanse of prairie or meadow – its keynote sounds made up from a distinctive blend of geology, climate and wildlife. In the wind, a forest sings. But a treeless plain nearby will, instead, vibrate like 'an enormous harp'. The Merrimack River murmurs, 'kissing' its banks while it flows. But a Swiss mountain brook will babble. And the Nile at Atbara will roar with fury.[8] At Coney Island in Brooklyn, the Atlantic waves come ashore on broad, open sandy beaches. They're gentle and slow. In the Azores, they hit the rocky shoreline with what has been described as 'a sharp, percussive, slaplike crack'. On the Suffolk coast in Britain, the steep rake of the beach means they arrive sounding agitated.[9]

And wherever we go, the heaving biomass of insects, birds and mammals creates its own tidal flow of noise, ebbing and flowing. Some creatures are at their most vocal when their habitat has dried out by the late-morning summer sun; others take advantage of the dewy dampness of dawn or late evening in autumn to project their calls across greater distances.[10] Go to parts of New Zealand or Australia and you'll be almost deafened by the cicadas. But only between December and March. In North America it's the massed croaking of frogs which announces the change of season.

Through sound, then, nature is our satnav, our clock, our calendar. And although we sometimes forget that today, when we have so much technology to help guide us, we would have been permanently tuned in to the natural soundscape for most of our past. As for our most distant ancestors, this soundscape would have been rich in significance – everything in it grabbing their full attention. It didn't just help them hunt down their prey in the darkest forests or show them when to sow their seeds or even provide their channel of communication with the invisible spirit world – though all this seems likely. The sounds of the wild also determined the first music our ancestors made, the first words they spoke. For the most important feature of early humans' relationship with nature is that they mimicked it.

We can still hear examples of this in the more remote areas of the world. More than thirty years ago, the American anthropologist Steve Feld noticed that the Kaluli people, who live in the rainforests of Papua New Guinea, have a wonderfully rich vocabulary relating to sounds. They have, for instance, completely different phrases for bird songs heard at ground level and those heard above, those heard nearby and those heard in the distance.[11] Show them a picture of a bird, and they'll imitate the sound before naming it. Ask them for the bird's name, and they won't tell you what it looks like; they'll say, 'It sounds like this'. Kaluli singing, too, is intricately connected with the everyday noises of the forest: birds, mammals, insects, trees, water flowing, rain pouring – these are what the Kaluli sing with, and to, and about.[12] The whooping and whistling and singing of the vocalists 'interlocks' with the ambient noise and observes its rhythm; it copies the insects and frogs in the nearby bush, and echoes back to them.

The Kaluli are not alone in having such acute powers of hearing and mimicry. In the Malaysian rainforest, we might hear the healing dance of the Temiar people, in which bamboo tubes are stamped on the ground in a pulsing echo of the cicada.[13] Or on the open grasslands of North America, the Native people of the Plains – the Blackfoot and the Sioux – who once hunted bison by driving herds into a ditch, might be heard singing, cunningly, in voices eerily like that of bleating calves. Though thousands of miles apart, both these groups of people have long thought of themselves as having come from the land – of having a kinship with its wildlife. They take the sounds of nature, amplify them, and return them to the world in an attempt to influence it.

The Sioux, the Temiar and the Kaluli provide an acoustic flashback to our distant past. In prehistory, to be human was to be a hunter-gatherer living in the forests or the open savannahs. Tens of thousands of years ago, as we moved stealthily through the undergrowth in search of food, information about the tracks and trails of particular animals, or news of any sightings, would have had to be communicated with each other. Like modern-day hunters in the bush, we would almost certainly have done this by mimicking the animals – their walk, their movement, their gestures, their *cries*.

Gradually, this mimicry formed into the words we uttered. It's why animal names, even today, are often onomatopoeic. They are made up of individual units of sound that capture the essence of the animals themselves, or which resulted from the gestures made by our tongues and lips when mimicking the creature. Over time, these evolved into more generic linguistic habits that we all share. Try this experiment, first done in the 1920s. Ask yourself which nonsense word would better describe the larger of two wooden tables: *mil* or *mal*? My guess is that you opted for *mal* – because that is what most people do, regardless of nationality. In lots of languages we associate the 'i' sound with small and the 'uoa' sound with bigger things. There have been more recent studies of the way animal names in many parts of the world capture qualities other than size – such as the fast-moving, twitchy essence of birds, or the slow, flowing movement of fish. So again, try this experiment. Which of these two words is a bird and which is a fish: *chunchutkit* and *mauts*? Again, my guess is that you got it right. *Chunchutkit* just sounds like a bird – or rather, it sounds like a bird *sounds*.[14]

It's in Latin America, however, that we're best able to see how deeply enmeshed past human cultures have been with the sounds of nature. People here didn't just engage in mimicry. In some of the ancient civilisations of Mexico or the Andes, whole symbolic belief systems evolved which put sound right at their centre. In the mountains of Peru, for instance, the Incas of the fourteenth and fifteenth centuries seemed to view the cosmos as being like a noisy pot-drum. The Quechua language of the region, even today, is full of words associated with containers or pouring, and these are also linked with natural phenomena. The earth, the sky, lakes, mountains, stones, houses and human beings: these are all viewed as being like vessels to be filled or poured out. When there's thunder, the saying in the Andes is that it's like a crack in the sky – in other words, like a break in a pot. Liquids – rain, urine, blood – all have fertilising power, so they were almost certainly fundamental to life and beliefs during the Inca period, and probably before. Containing or releasing them was most likely a sacred act, and pots would have reflected this. Sometimes they'd be beaten, so that their resonance could ring out like thunder. Lots of rituals evidently involved the sounds of pouring, with pots used to echo the natural world's soundscape of trickling rain or running streams, of urination, or perhaps bloodletting.[15]

Further north, there was an even dizzier array of sound symbolism. Archaeologists working on houses more than a thousand years old in Oaxaca on the south-west coast of Mexico have unearthed among the remains lots of bells, rattles, flutes and whistles. Many of the pots used for everyday cooking or eating have little hollowed-out feet which contained tiny clay balls acting like rattles. It seems people wanted, quite deliberately, to make quite a noise each time food was prepared or served. Farmers certainly wanted to kick up a din during fertility rites, since they used sticks with rattles to conjure up what the Aztecs thought of as the sound of a serpent. Clothing was embellished with beads or small bells, which must have jangled and chimed as people walked about. Some of these clothing bells had animal images appliquéd on to their tops – birds especially. Indeed, birds keep cropping up in the archaeologists' finds. Many of the pots are decorated with them, and many of the whistles and flutes are shaped like them.[16] In ancient Mexico, as in lots of other ancient cultures, birds clearly had great symbolic power. They were one of the many animal spirit guides, connecting humans with the hidden world of their dead ancestors. Any human-made item that captured some of their essential properties would also surely have recreated some of their valuable powers of sacred communication.

So it's not hard to imagine all these pots and whistles and bells being used in some noisy Aztec domestic ritual to evoke the spirit world by means of connecting their users with the natural world. And although we can't be sure exactly how far back in time these traditions go, we can guess they have their roots far back in prehistory. Aztec imagery makes a great deal of sound and speech and hearing. In their pictures, 'scrolls', decorated with precious objects, often emerge from the mouth of a speaker, as if the human voice were channelling the fragrant, flowery world of the gods and ancestors. But older stone carvings and murals also used flowers and swirls to represent breathing, speaking, singing, rumbling or echoing. Sound mattered in people's lives and beliefs enough for them to have tried hard to symbolise it visually for many generations.[17]

At no stage was this some abstract notion of sound. It was nearly always one inspired by the real world – by the singing wilderness of nature. In this respect, humans were just one part of a larger spectrum of creatures adding their own layers of noise to the bedrock of sounds caused by wind, sea, thunder and rain. But humans were never content just to copy the sounds of nature to help them get hold of their next meal. Culture kept evolving, giving us bigger, more complex ideas about what could be done with sound: how it could be shaped and manipulated to create dramatic effects and to help us understand our place in the cosmos; how nature might be not merely copied, but mastered.

4

A Ritual Soundscape

Most of us in Britain probably think of the Orkney Islands as rather exotic and remote: over 530 miles from London – more than 200 miles, even, from Edinburgh – and, although only ten miles north of John O'Groats at their southern end, always a flight or ferry ride away from the mainland. But in the Neolithic era, between 6,000 and 4,000 years ago, Orkney was far from being on the periphery of human life in the British Isles. It was, if only briefly, one of Europe's centres of civilisation. Here, just a few metres from the North Atlantic waves, and buffeted by the westerly winds, lies Skara Brae, a stunningly preserved sunken village of walled houses, complete with hearths and stone furniture such as beds and dressers; just a few miles to the southeast, two striking stone circles, at Stenness and Brodgar; and, close by, Maeshowe, a massive chambered tomb where the setting sun sends a beam of light along a tiny passageway to illuminate its inner chamber on the winter solstice. All are set in a spectacular natural amphitheatre of hills and lochs and bays.

It's here, perhaps more than anywhere for thousands of miles around, that we can appreciate a fundamental shift having taken place in human history. By the Neolithic, our ancestors were no longer having to make do with the world as they found it, seeking shelter and spiritual inspiration in nature. They were creating their own monumental buildings. These were not just physical objects. They were almost certainly also the components of a complex ritual life – places capable (though, at first, perhaps unintentionally) of creating extraordinary multi-sensual experiences, and, most likely, special kinds of noise. Tens of thousands of years earlier, the men, women and children of the Palaeolithic would have experienced the natural acoustic qualities of caves or canyons or forests. The Orcadians of the Neolithic, though, were putting up stone buildings of their own design – megaliths which turned out to have their own completely new acoustic identity and which demonstrated a new human mastery over the natural soundscape.

Of course, we don't know whether the Neolithic monuments of Orkney – or anywhere else – were built especially in order to create new sound effects. It seems pretty unlikely. But we can be fairly certain that no one was living in any of the stone circles or at the chambered cairn, simply because the detritus of domestic occupation has never been found there.[1] These were out-of-the-ordinary places, where out-of-the-ordinary events would surely have taken place.

What were these like? Again, we just don't know. But the lesson of ethnography is that in most cultures, rituals are multi-sensual affairs; and when you enter Orkney's Neolithic sites they certainly feel a bit like theatre stages, capable of creating lighting effects or strange smells, and encouraging us to move through them in unfamiliar ways. If people using these spaces 4,000 or 5,000 years ago noticed that the monuments that enclosed them also just happened to create striking sound effects, they must have been tempted to exploit them. But exploit them *how*, exactly? Were these the kinds of places where our ancestors came to make a spectacular din – or places where they came in search of silence and sensory deprivation?

The first, tantalising clues that might help us answer this question can be found at the Ring of Brodgar. It's a stunning site: a large, perfectly formed circle about a hundred metres across, formed by twenty-seven standing stones, some as much as four metres tall – all surrounded by an even bigger circular ditch cut out of solid rock. Ditches like this are quite common at stone circles, but very occasionally they are accompanied by raised earthworks, which create even bigger embankments. And the taller the embankment, the more it acts as a kind of 'baffle', blocking the sounds of the surrounding landscape, creating a kind of sound 'shadow' inside the circle. At Brodgar, however, there's no raised embankment; the whole site is more open to the surrounding landscape. One archaeologist has even described it as giving the impression not of having been built so much as simply 'gathered from the land'.[2]

Yet when it was first created, there could have been, not twenty-seven, but some sixty stones standing here, which would have created a much more enclosed space. Not completely enclosed, but enough, probably, to create a slightly different acoustic as you entered. Even now, if you stand right at the centre, it's possible in calm weather to just about detect distinct echoes ricocheting around the inside of the circle if you clap or shout loudly enough.[3] The effect is even more pronounced if you beat a drum, when the echoes seem to come back at you from all around the circle. Move away from the centre and the effect lessens; move to just outside the circle, and the effect's lost completely. The Brodgar standing stones are fairly roughly hewn on both sides, while at Stonehenge the stones appear to have had much smoother surfaces on the sides that faced inwards. This might have been a purely decorative design feature, but smooth surfaces would certainly have had the effect of making the standing stones at Stonehenge even better at bouncing sound back to the centre. So part of the sense of occasion in being at stone circles in the Neolithic era during some ritual could easily have been an acoustic experience, where there was a theatrical shift as you entered the circle – from the ambient sounds of the natural world, the sounds of the sea or of the wind, say, to a range of more contained and controlled ceremonial sounds within.

At the chambered cairn at Maeshowe, the effect is even more dramatic. The burial chamber itself is made of drystone walling and large flat slabs, but the whole structure is covered with a mound of clay and a layer of grass.[4] This provides effective soundproofing: very little in the way of noise can either escape from the chamber or seep into it from the outside. Yet, it would be misleading to assume that a cairn like this created a permanent ring of total silence in the landscape immediately around it, punctuated only by the odd seagull cry. For a start, the cairn at Maeshowe and others like it don't appear to be places where the dead were just left and forgotten. It seems more likely that they were entered fairly frequently, as if the ancestral remains held inside were themselves being visited, even taken outside, perhaps to take part in some ceremonial activity involving the living. It appears there was a fair amount of coming and going – *some* sort of activity regularly taking place inside.[5] And, if so, at least a bit of this would have been heard, even outside, since no cairn was ever 100 per cent soundproof. If sound leaked out, though, it did so in curious ways. Experiments conducted at sites in Ireland and further south in Scotland show that instead of *all* sound being blocked by the stone, it's mostly the high-frequency sounds that are trapped. Low-frequency noise, on the other hand, can escape. So if you were to do something like drumming inside the cairn, the normal beats you were generating would be transformed: someone standing outside at the back and sides of the tomb would hear something much deeper – indeed, something strangely deep, because it would sound as if it were rising up from the ground itself rather than coming from the tomb.[6]

Thousands of years ago, those people who were gathered outside these megaliths when others were busy inside would therefore have heard something. But it would have been something distorted, confusing maybe, a bit second-rate perhaps, compared with what was probably being experienced right at the hidden centre of these chambers.

Here, deep within these megaliths, was something new in human history: a handcrafted space of silence. Though it's possible that it wouldn't have remained silent for long. It would, for instance, have been easy to use this artificial silence as a blank canvas on which to create completely new soundscapes. The drystone walls of these cairns are smooth and precisely fitted together: perfect for reflecting sound efficiently. More specifically, their smooth, straight walls are perfectly shaped for creating what is called a 'standing wave' of sound, where sound folds back upon itself and builds up in complex layers. Start to speak more loudly, shout or sing in these precision-engineered cavities and it really doesn't take long before a rumbling effect is unleashed, which builds up in volume and intensity.[7]

So these can be quite noisy places. Yet, what if, instead of setting out to create acoustic chaos, the Neolithic people who might have gathered here used the artificial resonance to create a rather more controlled effect? It has been discovered that if we start beating a single drum at about two beats per second at the centre of these burial chambers, the timing of the echoes is just enough to trigger a feedback loop that is not only loud but also, in time, rather hypnotic. Some archaeologists have wondered about this hypnotic quality. They have noticed that the tonal frequency being created just happens to coincide with the same frequency that can send the human brain into a state roughly halfway between wakefulness and sleep – the state that prompts our strongest moments of vivid mental imagery and hallucination.[8]

At the same time, something else is generated that we can't actually notice: sound that exists below the threshold of human hearing. This so-called 'infrasound' affects us, too. It can make us feel pressure in the middle ear, give us a slight headache, even make us a little wobbly on our feet. A few years ago, at the Camster Round burial chamber in Caithness, on the Scottish mainland not far from Orkney, some volunteers were exposed to short bursts of infrasound created by drumming. They reported dizziness and, in some cases, a strange feeling of rising physically. Again, we don't know if any of this was an intentional part of the original Neolithic builders' designs. But if you had been in charge of some ritual taking place in here, it must surely have been tempting to heighten its other-worldly quality by doing whatever it took to provoke transcendental experiences.[9]

So far, we've been assuming that burial chambers like Maeshowe were places where Neolithic people didn't just leave their dead in peace and quiet but joined them at intervals for noisy, perhaps even mind-altering, rituals. But there's another possibility: that these were places where silence always reigned – even when they were being visited; that instead of people coming here to rave in groups, they came to be utterly alone, to be on some kind of spiritual retreat. The evidence for this possibility comes, not from Orkney some 4,000 or 5,000 years ago, but from France in the Middle Ages.

Near the village of Conques, in the old French province of Rouergue, there's a fascinating legend, all but gone from popular memory and surviving only in the most obscure references in a few old library books. Two giants were killed, so the legend goes, and their bodies laid out in a prehistoric tomb tucked away somewhere nearby. After lying in this tomb for many days, the two giants suddenly rose from the dead and were filled with dream-like visions of the future. They had, somehow, been reborn. There are more famous resurrection stories, of course. But what is interesting about this one is that very similar accounts, set in this same village of Conques, crop up in a famous eleventh-century manuscript, the *Book of Miracles of St Foy*, written by Bernard of Angers. These accounts refer to the ritual practice of incubation – people seeking a refuge in order to hibernate, and perhaps be renewed. And, indeed, people used to come to Conques regularly until the late fourteenth century, simply in order to lie for many days in or nearby the church sanctuary, in the hope of receiving a cure from the local saint, who would appear to them in some sort of dream vision.[10] As it happens, there are megalithic remains nearby. So it seems perfectly possible that this Christian practice took hold so vigorously here because it resonated so well with pagan practices that might well have been taking place in one form or another in this area for centuries.

The Italian researcher Francesco Benozzo has offered us a few tantalising scraps of evidence that support this idea of tombs as places of incubation – from various dialect names and other bits of speech. In lots of European languages, he suggests, the word for 'cave' comes from a common root meaning 'to hide'. Later, in Latin, this becomes something that means a 'cell'. In Old High German, the related word could be translated as 'house of dead people'. In Old Irish, we would translate the nearest word as meaning both 'hiding place' and 'silence'. In Old Welsh, it would be 'dream'. In other words, the various meanings spinning out from our one starting point link together the idea of darkness, a hiding place, silence and dreaming.[11] It all hints at a much gentler, quieter, more individual use for a sheltered space cut off from the outside world.

So is *this* what happened back in Orkney, in the inner chamber of Maeshowe? People crawled into it, not to bang and chant, but to rest and dream? Perhaps. Orkney was a part of the British Isles that fell under Norse control much later in its history. And we know that in some early North European cultures, young men were thought to need a period of dormancy before they stepped out into manhood – something that often entailed lying about for months near the hearth in a kind of ritualised lethargy, an extreme version of teenagers hibernating in their bedrooms.[12] So we can certainly imagine something like the 'incubation' practices of medieval France also taking place in the north of Europe – though in this case further back in time.

We'll never know the full story – whether the Neolithic monuments of Orkney, or indeed anywhere else, were places of noise or places of silence. What we do know is that they were places where human beings were beginning to discover that they could control the natural soundscape, and perhaps create wholly new ones of their own. It's a reasonable guess that the very act of controlling these new soundscapes – or at least being in charge of the rituals that may have invoked them – was hugely important. Not all and sundry would have been allowed inside a tomb like the one at Maeshowe. You could probably get about twenty people into its inner chamber at any one time. So who got in, and who didn't – who got to experience directly the sounds, smells and sights of whatever rituals went on there, and who got to hear just the muffled, mysterious rumbles outside – this was something that, if it really did happen, would have reflected, and reinforced, social hierarchies. And while we have no idea whether there were men or women in Neolithic Orkney who acted like priests – with the knowledge and skills to manage the ritual use of these spectacular sites, we can guess that there would have been people a bit like them; mystical figures, perhaps, who controlled sound in all sorts of wonderful – and secretive – ways.

5

The Rise of the Shamans

Strictly speaking, the shaman belongs to central Asia, specifically to the reindeer herders of Siberia and parts of China and Korea. Yet shamans – men, and sometimes women, who enter into a trance-like state in an attempt to contact the spirit world – can still be found in all parts of the globe. And historically, this special caste of people, who use either drugs or rhythmic music, or both, in order to communicate with the spirits, and then return to their senses in order to heal the sick, control animals or change the weather, have been found in one form or another wherever there have been hunter-gatherer societies, from the Native Americans of California, to the Inuit of Alaska, the San of South Africa and the peoples of Inner Mongolia. The practice of achieving some kind of ecstatic state is even more widespread. One survey of nearly 500 different cultures showed at least 90 per cent had what we might call, very broadly, 'shamanistic' features.[1]

This is significant if we are trying to understand the role of sound towards the end of the prehistoric era. In the late Palaeolithic and the Neolithic – that is, between about 20,000 and 4,000 years ago – most humans, like the reindeer herders of Siberia today, were hunter-gatherers. So it's not unreasonable to imagine that if our prehistoric ancestors were beginning to develop a religious life of ceremonies and rituals, there would have been among them a select group of people at least a little like shamans – people with a central role in such affairs. When we look closely at living shamans in action, we can sense how their drumming and their incantations, often in darkened spaces, would have comfortably fitted a Stone Age society dominated by a rich array of supernatural beliefs. So one obvious question is this: were the ways in which these shaman-like people used sound a means of wielding power and influence over their fellow human beings? To put it another way: how much was sound – a great force for social bonding, as previously discussed – starting to be used by skilled practitioners as a way of marking out and reinforcing new social and cultural divisions, or even hierarchies?

One place to start is for us to look more closely at what happens when a shaman enters a trance. Of course, no one shamanic ritual is quite like another. But one detailed account, now over a hundred years old, has become a classic. It comes from Waldemar Bogoras, a young Russian writer who had been exiled to Siberia for his progressive views by the autocratic Tsarist regime. Bogoras made the most of his lonely predicament by devoting himself to the study of the native peoples there – especially to the Chukchi, who herded reindeer across a large swathe of Russia's most north-easterly land. He was fascinated by their language, their lifestyle and, perhaps most of all, their religion. He quickly realised how important the shaman was to the Chukchi – and how skilled the shaman could be in deploying sound, not just to invoke the spirits, but also to create a richly suggestive atmosphere. He recounts, for instance, one evening that he had spent eating at a Chukchi camp:

> After the evening meal is finished and the kettles and trays are removed to the outer tent, all the people who wish to be present at the séance enter the inner room, which is carefully closed for the night.

When everyone is inside the small inner room, the lamps are extinguished, and the shaman lights his pungent tobacco and begins to operate.

> He beats the drum and sings his introductory tunes, at first in a low voice; then gradually his voice increases in volume, and soon it fills the small closed-up room with its violent clamour. The narrow walls resound in all directions. Moreover, the shaman uses his drum for modifying his voice, now placing it directly before his mouth, now turning it at an oblique angle, and all the time beating it violently. After a few minutes, all this noise begins to work strangely on the listeners, who are crouching down, squeezed together in a most uncomfortable position. They begin to lose the power to locate the source of the sounds; and, almost without any effort of imagination, the song and the drum seem to shift from corner to corner, or even to move about without having any definite place at all.

Before long, it's as if the room is alive with spirits, flying through the air:

> The 'separate voices' … come from all sides of the room, changing their place to the complete illusion of their

listeners. Some voices are at first faint, as if coming from afar; as they gradually approach, they increase in volume, and at last they rush into the room, pass through it and out, decreasing, and dying away in the remote distance. Other voices come from above, pass through the room and seem to go underground where they are heard as if from the depths of the earth.[2]

The effects this shaman achieved, with the help of some basic acoustical principles and by creating a suggestive atmosphere in that tiny, overheated space, are really striking. In the cold light of day, his ventriloquism would probably have been detected and lost some of its power. Bogoras spoke to Chukchi people who said they knew that the shaman's tricks were 'by no means real'. It was also understood that, like modernday spiritualists, they made generous use in the darkness of human assistants.[3] Yet, the Chukchi added, the shamanic performances were still 'very wonderful'.[4] Bogoras reckoned they worked on their audience because they weren't just random noises; they were sounds that made sense to a Chukchi beliefsystem in which everything in nature was considered as potentially alive and living, in which spirits were always lurking, ready to be released, in supposedly inanimate objects. Like shape-shifters, these spirits could take different physical forms. But they were most often invisible. They could only really be experienced through people hearing them.[5] So the shaman's small but significant manipulations of voice and drum in complete darkness would have provided those gathered round him with exactly the experience they would have expected.

In studying shamanism, we focus on the shaman because he or she is usually the one who enters the deep trance, who collapses and hallucinates and 'flies' among the spirits – who seems, in other words, to experience the other world most directly. But really, what is more interesting is the effect of all this on those who witness it. Most people, even today, don't need to experience full-scale transcendence or altered states of consciousness for themselves in order to feel religious. As the archaeologists David Lewis-Williams and David Pearce point out, 'They are willing to take the experience of others as indicative of a supernatural realm and to concern themselves with religious belief and practice'.[6] It's interesting that among the Chukchi, during a family ritual for slaughtering reindeer, Waldemar Bogoras noticed that everyone – not just the sha- mans, but ordinary men, women and children – would beat drums to 'call the "spirits" and try to induce them to enter their bodies':

> They *imitate* shamans, utter the cries of various animals,
> and make the peculiar noise supposed to be characteristic of
> 'spirits,' which is produced by a vibrating motion of the lips
> while the head is shaken violently.[7]

At other times, families acted together, shouting at the tops of their voices, to drive away evil spirits.

So shamans don't have a monopoly of spiritual power. What they do have, though, is equally significant: they provide everyone else with the model for how belief is to be acted out and sustained. They retain their influence by showing off their possession of special knowledge – including, of course, the ability to 'throw' their voice in the dark. It's why the Chukchi regularly hold competitions in which different shamans try to prove who's best at unleashing spirits, or controlling them.[8] Social influence – and, frankly, a good living – cleaves most easily to those who wow their audience the best.

We have to be really careful, however, about projecting this kind of behaviour back into prehistory and assuming that such things worked in exactly the same way then. The anthropo- logical study of present-day hunter-gatherer societies gives us clues to the past, but no certainty. Even so, when similar phenomena keep cropping up in different parts of the world, it suggests some degree of shared experience further back in time. And there are elements of shamanism that are undoubtedly universal features of religious belief. East or west, north or south, there was almost certainly a common need for our ancestors to make some sense of the dreaming or hallucinating they experienced for themselves, or witnessed in others. Sound as the means of accessing a spiritual or divine world has also been a remarkably pervasive feature of human culture. Over time, naturally, we came to experience these things in terms of our own particular culture and value system – a slow, complex process that has been called the 'domestication of trance'.[9] But even thousands of years later, when such rituals had already diverged into wholly separate religions, and the skills needed to manage these displays had become more institutionalised, we might still, with a little imagination, see the living remnants of much earlier practices.

A lovely example of acoustic trickery which bears an uncanny similarity to the shaman's practice of 'throwing voices' comes to us from medieval Christendom, and is still performed as part of an elaborate ritual in the English cathedral city of Wells. Nowadays, the effect is charming rather than awe-inspiring. But it still gives us a hint of how, in the thirteenth century, when the ritual began in an organised way, ordinary God-fearing men, women and children might have been led to believe they were hearing something supernatural.

Once a year, in spring, the congregation gather just outside the cathedral's spectacular façade: Wells's West Front. The cathedral itself isn't very big – its nave is only about half the width of, say, Notre Dame in Paris. But its West Front outside offers a façade that is almost as wide as its French cousin, and is covered from top to bottom with around 300 statues – of kings, saints, apostles, bishops, angels, and, right at the apex, Christ himself.[10] The effect is stunning, and would have been even more so when it was first completed nearly 800 years ago, when the statues and niches would have been painted in vivid colours – red, green, and gold especially. It's been likened to a giant upright page from an illuminated manuscript.

This extraordinary feast for the eyes is spectacular enough in itself. But on Palm Sunday, the façade suddenly reveals a hidden power: it bursts into life through sound. As the congregation assembles quietly outside, a procession of the cathedral's choir and clergy stops next to them and begins singing the Introit from Psalm 24:

> The earth is the Lord's and all that fills it,
> the compass of the world and all who dwell therein.
> For he has founded it upon the seas
> and set it firm upon the rivers of the deep.

The next line will then seem to come, not from any of the choristers standing right in front of the congregation, but from what sounds like a disembodied voice, floating down from somewhere up on the façade itself:

> Who shall ascend the hill of the Lord,
> or who can rise up in this holy place?

It's mysterious. None of the congregation standing on the ground outside can *see* anyone up on the façade. The ethereal voice seems, in fact, to be coming from one of the angels that are set in niches about a third of the way up the West Front. It's as though this particular angel were singing to the congregation – as though the angel itself were alive.

It's a wonderful auditory illusion. And the secret behind the singing angel is only revealed if you enter the cathedral itself. Inside, just to the right of the main doorway, you'll find a rather nondescript and easily overlooked wooden door. It's usually locked, but if you can arrange for it to be opened, it reveals a rabbit warren of staircases and corridors hidden between the outer wall of the façade and the interior wall of the cathedral's nave. One of these corridors eventually leads you to a narrow, somewhat chilly gallery, which runs just behind the statues of our singing angel. The reason for the slight chill quickly becomes apparent. Arranged in a line along the outer wall of the gallery are several large holes, or *occuli*, through which the cool outside air enters. They are clearly part of the original design, and their function is revealed by their rather unusual shape. The occuli are like giant cones, larger on the outer wall of the cathedral than on the inner wall of this hidden gallery; in other words, they're shaped like megaphones, designed very clearly to amplify any sound made inside the gallery. Some of the occuli are set about four feet above the gallery floor, while others are positioned about five and a half feet up – just at head height for either a child or an adult chorister. And, of course, it's precisely here, during the rituals of Palm Sunday and Easter, that one or more choristers will have been secretly positioned, to sing into these very portals – and to create the illusion for those standing on the ground outside of angels in full voice.

Even higher up, there are other hidden galleries where the West Front is pierced by yet more occuli, this time too deep for singing but positioned directly behind a row of statues of trumpeting angels. It's quite likely that these, too, were used by human trumpeters to announce the arrival of the Palm Sunday procession, surreptitiously recreating those angels sounding their trumpets at Christ's entry into Jerusalem. Indeed, we can begin to imagine the whole West Front as being a scaled-up musical instrument, emitting all sorts of wondrous sounds at high volume.

Of course, as modern-day sceptics, we're hardly shocked at the acoustic trickery of Wells Cathedral. We never believed in the first place – did we? – that a stone statue of an angel could actually sing. Most of us don't even believe in angels.

Yet in the thirteenth century, belief in miracles was a normal, unquestioned part of everyday life. And the singing and trumpeting angels of the brilliantly coloured and glittering West Front would have made sense in a visceral kind of way to the deeply superstitious people gathered below. They would have heard about the Heavenly City of God and of Christ's entry into Jerusalem. And at Wells, as the procession moved under the singing and trumpeting angels, they could easily imagine what it would be like when it was their turn to make that journey from one world into the next.

The charm of this performance today should not blind us to some hard power politics behind its creation in the thirteenth century. Then, the singing angels brought kudos to a cathedral church with no major relics of its own and still very much in the shadow of its neighbour, Bath Abbey. The illusion did exactly what the bishop of the time, Bishop Jocelyn, and his architect, Adam Lock, had set out to achieve: it offered a spectacular attraction that heightened the church's grandeur and magnified its power to protect the faithful. But, more fundamentally, the sound of the Palm Sunday performance brought a clear reminder of the Church's power to make things happen. Its intended audience wasn't lectured about salvation, it was allowed to feel as if it had experienced it. Like all good spectacles, then, the Palm Sunday ritual at Wells worked at the emotional as well as the intellectual level. Its arresting sound was tangible enough to be interpreted as genuine by those present – though also ephemeral enough, and of course invisible enough, to escape proper scrutiny. In other words, it was perfect for those seeking to deploy the remarkable power of suggestion.

The special status granted to those who could pull off such displays was hardly unique to medieval Christianity. It's likely that for centuries control over sound had marked people everywhere as special. Spirits and gods were widely conceived as invisible but audible things – as sound, wind, vibration. And it was readily accepted that not everyone could detect, let alone interpret, these subtle markers of presence. In central Asia, the earliest shamans almost certainly could. In some later Sufi Muslim traditions, so too could those who 'worked themselves into a mystical trance by chanting and whirling in slow gyrations'.[11] In the ancient Iranian religion of Zoroastrianism, it was the priest Srosh who represented the genius of hearing. It was Srosh who stood between humans and the pantheon of the gods, listening out for divine messages and communicating them to the rest of humanity.[12]

If we define ancient shamanistic practices fairly loosely like this, we should also define fairly loosely the social status of these early priest-like figures. In Siberia, Waldemar Bogoras discovered shamans who were more like plumbers or electricians on call, ready to drop by and do the rather grubby job of removing an unwanted spirit through a quick incantation delivered for a small fee. Yet there were obviously other grander figures around, too. Those who have studied South American shamanism suggest there might actually be two fundamental versions: one that is more democratic and inclusive; the other stressing the esoteric knowledge that is held by a small elite. If so, it's probably with this second kind – those who guarded their acoustic tricks most closely and used them to impress whole communities – that access to the spirits evolves into real social power.[13]

But this isn't just about the emergence of a religious ruling elite. In the first part of this book, I've stressed how the distant human past was probably far from silent – how, instead of treating ancient archaeological sites as places for hushed reverence, we should probably reintroduce a sense of the noise and life which they contained. Sound was a crucial component of the social bonding that drove human evolution – its rhythms, in particular, drew from nature and from our own bodies, and helped small communities work together. Yet as humans moved from making noise in the places they found in nature – the forests or the caves – and started to build ever more elaborate structures of their own, the rather more controlled and artificial noises made in these places would also have sown division. As societies became more complex, so music and performance became more complex, more specialised. The distinction between those who could make impressive sounds of all kinds and those who couldn't widened.

Music, like talking, was once a shared, socially bonding activity. Now, increasingly, such sounds would divide us in two: there would be performers and audiences, and these two groups often rubbed up against each other in an uneasy alliance. For, even in the world of oratory and theatre and imperial spectacle, being in an audience – supposedly just listening – was far from the passive activity that we might expect. Sound could certainly be deployed in the service of the powerful, but it was always too slippery a thing to be fully owned or controlled. It was something that could also be used to call rulers and public figures to account in surprising and dramatic ways.

II
The Age of
Oratory

6

Epic Tales

In 1933, a young classics scholar from Harvard University began a two-year journey by car and on foot through the scattered hill villages of southern Bosnia, Serbia, Albania and Montenegro. It was a Balkan journey that would dramatically turn upside down what historians thought they knew about the origins of the world's oldest literature. The scholar's name was Milman Parry, and he was on a hunt for accomplished local poets and singers. This wasn't as easy as it might at first appear: these weren't the kind of people who were ever going to be published, let alone famous; they would only be found, if at all, by word of mouth. The best bet, Parry discovered, was to visit Turkish coffee houses. On market days they would be full of peasants; in Muslim areas during Ramadan they were centres of evening entertainment for every villager.

One day, talking to the people busy smoking and drinking inside a small café tucked away on a side street, Parry and his young assistant Albert Lord were told about a singer who – everyone agreed – was the greatest performer in the district: an elderly peasant farmer called Avdo Mededović. Parry and Lord quickly tracked him down, and recorded him just as they found him: sitting in front of a small crowd playing cards, cross-legged on a bench, sawing away at a simple string instrument while swaying in time with the music and reciting his verses in a strange but mesmerising rhythm.

Avdo Mededović was good. But there were others like him. By the time Parry had finished travelling around and transcribing all these singers' lyrics, he had filled nearly 900 notebooks and created more than 3,500 double-sided aluminium discs of recordings. What they captured were the words and voices of hundreds of men and women who were unable to read yet somehow capable of reciting epic stories, many of which extended over not just hours but whole days. By 1935, Parry had seen and heard enough to reach a firm conclusion. Singers like Avdo Mededović, he decided, were able to perform these Serbo-Croatian epics because they didn't really have to memorise them wholesale: they assembled them as they performed them, using a range of stock phrases with which they were already familiar and which would ensure the lines always somehow fitted the required metre and rhythm. So there would be lots of repetition and lots of formulas: the settling of the morning dew would always be recited as 'The chilly dew fell', every garden was a 'green garden', hazelwood would always be 'young', and so on. Naturally, each time an epic poem was recited it would come out a little differently, but its essence would live on.

Milman Parry and Albert Lord hadn't, of course, discovered oral poetry. Anthropologists in the 1930s already knew of 'unlettered' people with extraordinary literary talent in other parts of the world. There was the Epic of Manas, which had been recited among the Kyrgyz of central Asia for at least two centuries; and the Mande Epic of Sunjata, which had recounted stories of battles and heroism in West Africa since the thirteenth century. There was that great Sanskrit epic the Mahābhārata, from ancient India, and the equally famous Anglo-Saxon story of Beowulf. All these, and more, were known as stories which, for hundreds of years, had been spoken aloud rather than written down. But what Parry and Lord offered for the first time was a convincing account of exactly how such monumental stories could have been composed without a script. This was revolutionary because it allowed the two men to reinterpret radically the origins of the oldest, most famous epic in all of western literature: the *Iliad*.

One of the earliest known copies of the *Iliad* is held at the Marciana Library in Venice. Although there are fragments of it on bits of papyrus that are even older, the rare and exquisite manuscript that is housed here is the one on which most modern editions of the vast, heroic story of Achilles and the siege of Troy are based. Yet even this *Iliad* isn't really original. It's in a form of ancient Greek, but it's medieval – handwritten some time in the tenth century AD. The epic itself has probably been performed since about the seventh century BC. Even then, according to some, it was an Iron Age retelling of stories already several centuries old – which, if true, would take us back to a point before the arrival in ancient Greece of the alphabet.

Its deepest origins, probably from somewhere on the west coast of what is now Turkey, have vanished into thin air. Parry and Lord, however, gave us a theory to explain its long and hitherto obscure gestation through the ephemeral medium of the spoken word. Just as the poets of the Balkans had used their stock phrases and musical rhythms to hold together their own recitals without having to write things down, so too had the distant creator of the *Iliad*: a person – or, most likely, several people – whom we later refer to as 'Homer'. Time and time again, we hear in the *Iliad* about 'swift-footed Achilles', the 'rosy-fingered dawn', the 'wine-dark sea', the 'bronze-clad' Trojans. These epithets were all, surely, remnants of the epic's original building blocks: set pieces and set themes that were assembled and reassembled ad hoc to create painlessly the required metre and rhythm during each live performance.[1] In other words, the form and the content of the *Iliad* so neatly echoed the Balkan recordings that it, too, must surely have been forged not on the page but through speech. And if a single 'master poet' did step in at some point to give it a little extra coherence, that must only have happened after a long, long period of improvisation.

All of which turns one of the greatest works of literature in history into a piece of early sound art. With what appears to be a rich layering of several archaic Greek languages, the *Iliad* was, so the argument goes, the sublime reflection of an incredibly rich oral culture that existed in Greece and across much of the ancient world.[2] This was a world on the brink of literacy, but where language was still more a matter of speaking and listening than writing and reading.

For a history of sound, then, epic tales of the ancient world like the *Iliad* are crucial. They remind us that civilisation can be sculpted not only from the written or printed page, but also from something as democratic and free as the noise of human speech. And we certainly need to value cultures that have too often been judged by literate, western standards, then side-lined as 'primitive' and lacking.

Yet, a note of caution. The whole idea of oral culture is so appealing that it can sometimes be invested with a little too much romance. We can be seduced into seeing it as an inno-cent, pure, natural, unchanging thing, part of an authentic folk life – idealising it, perhaps, in a reaction to the artifice and technology of our own modern world. History, though, muddies the water. For a start, there's evidence that the ancient world was far from being wholly illiterate. There had been manuscripts in places like Mesopotamia for more than a millennium before the *Iliad* first appeared. The Sumerian *Epic of Gilgamesh* was written in a form of ancient Babylonian around 2000 BC – some 1,300 years before Homer's epic tale. By this time, largely on the back of extensive trading by the Phoenicians, an alphabetic script was spreading around the Mediterranean, and being adapted to various local languages; other scripts were taking off in Pakistan, China and South America.[3] By the late sixth century BC, in Greece alone there were some 200,000 epic and lyric verses, or works of prose, circulating in writing – a century later, there were encyclopaedias, travelogues, histories and inscriptions everywhere. So ancient Greece was neither fully oral, nor fully literate: it was a complex, messy mixture of both.[4] As for the *Iliad*, many scholars insist that it's just as likely to have been a patchwork of written sources from several authors, or even the scripted creation of a single poet – someone who over a period of decades gradually enlarged a small poem, 'composing new bits and literally cutting and pasting them into the original papyrus roll'.[5]

So the *Iliad* might have been a written masterpiece after all. We just don't know. What we do know is that in ancient Greece the spoken word still carried far, far more power and influence than it does today, which means we have to think about what kind of cultural life this created. We know, for instance, that what can't be written down has to be remembered, and what has to be remembered needs to be memorable. All sorts of information of communal benefit have to be constantly repeated if they are not to be lost for ever. The Yupik people of south-west Alaska supply living proof of this. They have berry-picking songs that store information about likely spots for finding this food. They also have travelling songs packed with details of the weather conditions you should expect on various routes across the ice.[6] It's likely that ancient Greeks had their own mnemonics like these. They would also have had the rituals and simple routines of daily life, which stored and kept alive knowledge: how to name the young, or heal the sick, or bury the dead.[7] It's why rituals have always been important in pre-literate societies – and why, in classical Greece, the philosopher Plato worried about what would happen if everyone was suddenly able to write things down:

> If men learn [writing], it will implant forgetfulness in their souls: they will cease to exercise memory because they rely on that which is written, calling things to remembrance no longer from within themselves.[8]

Plato was pretty clear about where *he* stood. But from today's vantage point, it's harder to be quite so clear about the merits of an ancient world dominated by speech. Some modern writers see its highly ritualistic nature as having stopped intellectual development in its tracks, arguing that it left no room for breaking out of old habits or formulaic ways of thinking. The arrival of writing, they suggest, was revolutionary because it allowed ideas to be transported long distances and circulated more widely. Its arrival even freed people to think in new, abstract ways – to accumulate knowledge and become philosophers or scientists.[9] Others, though, remind us that for many centuries writing and reading were skills enjoyed by only a small elite. In which case, they argue, literacy restricted the spread of ideas and information, and magnified the powers of a small priestly or princely caste. Speech, they point out, had always been understood by anyone within hearing distance brought up in the same language.[10]

Even as the alphabet extended its grip, speech was being defended for its direct, unmediated quality. In the ancient Hindu tradition, for example, the spoken words of the Veda long remained a protected focus of religious life. Hindus insisted that a mantra had to be voiced correctly, because the sacred was made manifest through sound: if the sound wasn't right, its effectiveness would be undermined; if it was right, it could overpower and cleanse any sinner who was forced to listen. Composed between about 1500 BC and 700 BC, the Vedas were first recited in a completely oral society, and even centuries later, books were treated in Hinduism as 'graven images'. As far as the Mahābhārata was concerned, those who wrote down the Veda would 'go to hell'. Its words were things to be communicated in person. That way, apart from anything else, Hindus could have direct contact with a living guide.[11] Speech was always understood as something intimate, alive, sociable.

Frankly, the whole idea of two competing worlds of speech and writing locked in a struggle for dominance in the ancient world speaks only to the anxiety that is always experienced when a new and unfamiliar medium arrives. In reality, writing didn't destroy oral culture; it just absorbed some of its essential features. In Hinduism, the sense of the sacred that was originally attached to the speaking of holy words was gradually inherited by holy books, so that their presence as physical objects came to have the same powerful effect. These holy books, of course, continued to be read aloud – to be listened to – in thousands of temples or churches. Books of all kinds continued to be read aloud everywhere into the Middle Ages and beyond. For centuries, universities would examine their students through verbal interrogation. Oaths and wills would be legally binding through being uttered. And later still, some of the aesthetic traditions of ancient epic poetry, with its improvisation around stock phrases, would find an echo in musical traditions such as ballads or jazz or rap.

Perhaps the main reason oral culture survived the first shock of writing is that it wasn't just a one-way medium. When epic poems were recited, it was always in front of other people.

Speech wasn't something hurled at audiences on a take-it-or-leave-it basis. It was woven together with their active involvement. Indeed, speakers found themselves having to bend to their audiences even in the most formal and spectacular of settings.

Take the magnificent ancient theatre at Epidaurus in southern Greece, for example. With its beautifully symmetrical, gently sloping auditorium curled around more than half of the flat, circular stage below, it forms a vast bowl cut into the landscape. It's not quite on the scale of the Hollywood Bowl in Los Angeles, but it's not far off – and it's clearly one of the inspirations for that particular 1920s creation. When Epidaurus was first used to stage theatrical productions back in the third and fourth centuries BC, up to 14,000 people could be seated there comfortably.[12] It's easy to imagine that if you were one of those sitting in the back row near the very top of the auditorium, it would be all but impossible to either see or hear what was going on below. However, in classical theatre, the distance of the actors – and the fact that most of them wore masks – would have meant that spoken language had to do most of the work, so it's just as well that what the people down below were saying could be caught with stunning clarity.

The secret of Epidaurus' success lies mostly in the fact that it's an open-air building, which cuts out the risk of any interfering reverberation. But it's the perfectly sloping auditorium, with its rows of limestone seats, that really makes a difference. It allows voices from the stage below to somehow 'roll' upwards without bouncing around too much, with low-frequency rumblings being absorbed and the higher-frequency sounds of the actors and the chorus being amplified as they go.[13] The quality of the experience for those trying to hear something in a place such as Epidaurus was evidently taken very seriously. It's a vivid reminder of how in oral cultures listening to stories was just as important as telling them.

It's therefore no surprise to find that audiences had real power to shape what they heard. At these open-air theatres, instead of being concealed in darkness, the seated crowds could easily be seen by the actors – and they could easily be seen by each other. Instead of being forced into a hushed reverence, their attention focused on an illuminated stage, they could remain talkative and unruly. In the early days, when seats were made of wood, audiences would sometimes show their disapproval of anything going on below by drumming their heels against the benches. Plays could be disrupted by shouting, jeering, even throwing food. These shows were, after all, just one small element in a whole day's worth of festivities: it would have been hard to instil in the crowds a sense of reverence for some events and not others.[14] Theatre-going was something lots of people did – and they did it often enough not to be overawed by it – so the ancient Greeks were picky and assertive. Playwrights and actors had to work hard to grab their attention, to make sure they were spoken to, interacted with; the chorus, especially, would try to address the spectators directly, as if facing a jury. Their words were not just audible; they were aimed at an audience that knew how to listen. As historian of Greek theatre, Peter Arnott, points out, 'the public was an active partner, free to comment, to be commented upon, to assist, or to intervene'.[15] Audience participation was as much a part of a performance as were the actors in the orchestra or on the stage.

Of course, an epic tale of the kind we have been discussing was probably never performed before thousands of people in a theatre like Epidaurus. It was more likely to be recited in front of a small gaggle of people sitting around in the shade under a tree – or, as it was in the Balkans for Milman Parry and Albert Lord, next to a group of smoking, drinking card-players at a café. But the essential point remains. In a vigorous oral culture, it's hard to draw a firm line between speaker and listener.

Stories – and the great works of literature based on them, like the *Iliad* – emerged from a kind of dialogue between the two. When it comes to the human voice, especially, sound just can't be contained. Once uttered, it's too much 'in the world' to belong to one person. It's inherently public, open to scrutiny. And this, as we'll find out, had profound implications for those in the ancient world who sought to persuade the mass of people to support them and their ideas through the sheer power of speech.

7

Persuasion

When Barack Obama spoke on a damp, breezy night in Chicago's Grant Park in November 2008, having just won the election to be American president, his speech that evening seemed to confirm one thing: that whatever people in America or the rest of the world thought of his policies, here was a man who could talk brilliantly – better, certainly, than any of his political opponents; better, probably, than any president since Roosevelt even, some have said, since Abraham Lincoln. The speech in Grant Park proved – if proof were needed – that Obama's eloquence was one of his key election weapons. It persuaded his supporters to feel hopeful and his opponents to feel generous. It encouraged them all to believe in the same things that he believed in – and, perhaps, in the process, to heal some of the rifts that had opened up across America during the campaign. Like all the most powerful speeches, it was full of 'ringing truths and vital declarations'.[1] It also reverberated with echoes from the past, of emancipation and conciliation. Obama was the first African American to be president, yes. But here was someone who transcended race. Through his speech, he seemed to suggest that his victory embodied something bigger taking place in American society. He was inviting his audience to see that history was being made there and then – by them, as well as by him.

It's been said that Barack Obama's knack is that whenever he speaks he evokes all at once the spirit of Abraham Lincoln, Martin Luther King, Woody Guthrie and Sam Cooke.[2] It is partly a matter of what he says, and partly a matter of how he says it – of cadence as well as content. And the effect has been to conjure up for his audiences a 'live connection to American history'.[3] Yet there's another tradition that Obama's speeches have always invoked: the oratory of ancient Greece and Rome, where the art of public speaking was right at the heart of daily politics. It wasn't about crafting fine words for mere aesthetic pleasure. As we learned in the last chapter, when someone recites a story out loud, its success – indeed its very worth – depends on the reaction of those gathered around listening. So political oratory is about the art of persuasion, about changing minds. It's also therefore about the audience's skills. It's the listeners who have to interpret the words heard – and perhaps detect in a speaker's voice any falsehoods or manipulations designed just to get their vote. Which means that alongside the art of persuasion, there is also the art of auditioning. Understanding its origins takes us well beyond Greece and Rome – eastwards, to the first Buddhists and the beginnings of Islam.

Let's start, though, not with the listeners, or in the East, but with those making the speeches in Rome, some 2,000 years or so ago. For Romans, the art of oratory was entwined in everyday life – in law, in administration, in learning, in bartering, in rituals and ceremonies. Most Romans who wrote, wrote to speak and to be heard; they knew the power of the voice and how cadence could make or break a sentence. But it was in the Roman Forum where the big set-piece speeches would have been heard. Nowadays, it's an atmospherically ruined oasis amid the roar of Rome's traffic – quiet, except for the murmur and footsteps of tourists. During the ancient republic, however, it was the gleaming, monumental heart of public political life, the air filled with the sounds of passionate argument. This was for hundreds of years the most celebrated centre of oratory in the western world – just as Athens had been a few centuries earlier.

If I were a Roman citizen wanting to witness some of the best political speaking, the best place to come would probably have been the area in front of the senate house, near the ruins of the old comitium. It was in this part of the forum that speeches were made by the ruling elite, often facing out across the square from a raised platform – the so-called 'rostra'.[4] When people think of Barack Obama's powers as a speechmaker, they'll often compare him to one man in particular who performed regularly here – perhaps the outstanding, if ultimately unsuccessful, Roman politician of the late republic and certainly one of its greatest orators: Marcus Tullius Cicero. His speeches have gone down in history as so well structured, so dramatically delivered, so persuasive, that his name, even today, personifies eloquence. Cicero showed that in ancient Rome good oratory before the voting public here in the forum was the supreme political skill: you had no hope of gaining public office if you couldn't speak well. Yet speaking well could cover a multitude of sins, as well as a multitude of virtues. Right from the start the voting public would have known there was a very thin line between being persuaded and being cajoled, between being inspired and being bamboo-zled, between being thrilled and being misdirected.[5] So public speaking and public listening are not just part of the history of democracy; they are also part of the history of deceit.

What, then, were the tricks of the trade practised by the ancient Greek and Roman orators, which find an echo in the speeches of modern politicians such as Obama? 'Rhetoric' is full of rules – so many, indeed, that for centuries it ranked alongside grammar and logic as one of the core skills in a university education. Nevertheless, one of the classical devices that most often stands out in any politician's speech is the 'syntheton', in which two ideas are joined together, like, say, 'truth and justice'. Obama is particularly good at another, more subtle rhetorical device, the 'tricolon', in which ideas are unfurled in threes. The most famous example from Rome is not from Cicero, but Caesar: 'I came, I saw, I conquered.' As for the American president, recall these few lines from his November 2008 speech in Chicago:

> It's been a long time coming. But tonight, because of what we did on this day, in this election, at this defining moment, change has come to America.

And this, from an earlier speech – to the 2004 Democratic convention:

> Tonight, we gather to affirm the greatness of our nation, not because of the height of our skyscrapers, or the power of our military, or the size of our economy. Our pride is based on a very simple premise, summed up in a declaration made over two hundred years ago: 'We hold these truths to be self-evident, that all men are created equal, that they are endowed by their Creator with certain unalienable rights, that among these are life, liberty and the pursuit of happiness.'

This passage also reveals another nice rhetorical touch: drawing attention to a subject by circling around it rather than discussing it directly. Obama boasts about America's might, yet somehow without seeming to boast about it.[6]

Repetition, though, is what seems to infuse a speech with a feeling of soaring poetry. So if we look once more at Obama's November 2008 speech, we will find the device of 'anaphora' at the beginning of successive clauses:

> If there is anyone out there who still doubts that America is
> a place where all things are possible, who still wonders if the
> dream of our founders is alive in our time, who still questions
> the power of our democracy, tonight is your answer.

And, from 2008, the device of 'epiphora' at the end of sentences – including 'Yes we can', which was repeated five times in one section of his speech.

If all this talk of anaphora and epiphora makes rhetoric appear cold and mechanical rather than visionary, it's important to remember that a clear set of overarching ideals for good speaking also existed. The Greek philosopher Aristotle suggested that rhetoric should have pathos, logos and ethos – in other words, it needed to be a mix of emotion, argument and character. For Aristotle, then, rhetoric was not just about learning 'a series of tips and tricks for momentary success in speaking'; it represented 'a theory of human nature'.[7] And when Cicero spoke, he claimed there was also some higher goal in his mind – some civic ideal being embodied in what he said and how he said it. In speaking clearly, deliberately, rationally and thoughtfully, he hoped to propel the Roman state towards a calmer, wiser way of governing.

As for Obama, his speech also tries to convey something of his ethos through projecting something of his own, complicated identity. One way he does this is by talking constantly about the need to avoid narrow bipartisanship. The other way is through his voice and a very distinct style of delivery.

It's often been said that a speech doesn't come alive until it's spoken – and this is especially true of Obama. As the novelist Zadie Smith points out, Obama's voice is rarely unchanging and singular in style. Even in his autobiographical writing, he's 'many-voiced'. In his memoir, *Dreams from My Father*, for example, he shows that he has a good ear and an ability to turn it into dialogue. So he can 'do young Jewish male, black old lady from the South Side, white woman from Kansas, Kenyan elders, white Harvard nerds, black Columbia nerds, activist women, churchmen, security guards, bank tellers, and even a British man called Mr Wilkerson'. The result is that when Obama steps up to the podium, he 'doesn't just speak *for* his people. He can *speak* them.' [8]

His critics, of course, recoil at this. They suspect that when he slips from one kind of inflection or accent to another – like Tony Blair, dropping his aitches to switch from ex-public schoolboy to 'mockney' lad – Obama is betraying the fact that he lacks some kind of essential, unitary self. This, in turn, proves he's just another Janus-faced politician saying different things to different people. Where, they ask, is the real Obama? For, as Cicero said, the 'good speaker' should always play himself. Yet, as Zadie Smith argues, speaking in many tongues is Obama playing himself – demonstrating that he comes from a mix of races and cultural backgrounds – and that in having a complicated back-story he's no different from most of his audience. In weaving a strong sense of inclusion, in swapping the language of 'me' for the language of 'we', it's the voice itself that can be – and probably always has been – the most powerful tool in a speaker's kitbag.

No wonder, then, that in ancient Greece and Rome public speakers looked after their voices just as carefully as today's opera singers – or, indeed, just as carefully as the classical world's own star actors. As far as Plato was concerned, being 'fine-voiced' was the very definition of what acting was all

about.[9] So there was lots of dubious advice about how to pre-
pare the voice for performances. It was vital, for instance, to
avoid sex, to be physically robust and to have the right diet.
Roast meat, in particular, was to be scrupulously avoided.[10]
This tradition of voice-training was later adopted by the
Romans. In the pages of Suetonius we find Emperor Nero
– a keen amateur performer – lying on his back with lead
weights on his chest, abstaining from fruit and purging himself through laxatives and vomiting before striding on stage.[11]
It's an extreme example, and it's true that everyone thought
Nero was being absurd. But few disagreed with the idea that
on stage an actor's voice needed to be carefully modulated. A
Roman advice manual, *Ad Herennium*, points out the importance of controlling pace and varying tone:

> Pauses strengthen the voice. They also render the thoughts
> more clear-cut by separating them, and leave the hearer time
> to think. Relaxation from a continuous full tone conserves
> the voice, and the variety gives extreme pleasure to the hearer
> too, since now the conversational tone holds the attention
> and now the full voice rouses itself ...[12]

This quote also shows us very clearly how important the
audience was in all this. When Cicero delivered his first oration against his political opponent Catiline, accusing him of
plotting to overthrow the government of Rome, he addressed
much of the speech to Catiline himself, but also used the audience's embarrassed reaction to strike the killer blow:[13]

> You came a little while ago into the senate; in so numerous
> an assembly, who of so many friends and connections of
> yours saluted you? If this in the memory of man never
> happened to any one else, are you waiting for insults by word
> of mouth, when you are overwhelmed by the most irresistible
> condemnation of silence? Is it nothing that at your arrival

all those seats were vacated? That all the men of consular rank, who had often been marked out by you for slaughter, the very moment you sat down, left that part of the benches bare and vacant? With what feelings do you think you ought to bear this? On my honour, if my slaves feared me as all your fellow citizens fear you, I should think I must leave my house. Do not you think you should leave the city?[14]

Cicero's performance that day must have been mesmerising. But even to talk of his oration as a performance brings us up against the chief complaint against rhetoric: that it's all an act. And that if it's so obviously designed to work a spell on us, we should distrust it. Even in the fourth century BC, Plato was predicting a slippery slope where style triumphed over content. The desire to seduce an audience through technical virtuosity, he reckoned, would only encourage displays that were more and more histrionic. Speakers, for instance, might supply their own sound effects:

We shall have the noises of thunder and wind and hail, and of wheels and axles; the notes of trumpets, pipes, whistles, and every possible instrument; the barking of dogs, the bleating of sheep, and the twittering of birds. All these will be represented by voice and gesture, and narrative will play only a small part.[15]

Plato's objection wasn't aesthetic. He worried more about how easily the mob could be swayed. An orator, he believed, didn't need to know things; to succeed he just needed to find a way of persuading 'the ignorant' that he knew more than those who really *did* know. Plato always had it in for the mob, of course, but he wasn't alone in thinking of rhetoric as a form of witchcraft. In Aristophanes' play *The Clouds*, we find rhetoric being ridiculed mercilessly for allowing false arguments to triumph over strong ones. And, centuries later, we find St

Augustine urging 'good men' to take up the arts of rhetoric and fight those who used it 'to further injustice and error'.[16]

It is difficult to shake off the feeling that rhetoric served best the purposes of the demagogue. This is understandable, but also, perhaps, a little misplaced. For it overlooks another great tradition that existed in ancient civilisations alongside the training of speakers: the training of listeners to become as attentive as possible. It was a tradition in which listening, not speaking, was the key ethical act. And it was a tradition at its strongest to the east of Rome and Athens.

Take a walk through any of the big cities across the Middle East today, and when you come to a busy market, the swirl of chatter and music and nearby traffic can be overwhelming. But sometimes you can also just about make out through the fog of noise the faint strains of another sound spilling into the street – from loudspeakers in cafés, shops and workshops, or from the open windows of houses, apartments or even passing minicabs. It's the sound of CDs or cassette-tapes of sermons by the star preachers of Islam.

To listen to these recorded sermons is the contemporary version of a long tradition stretching back to the foundational years of Islam in the seventh century. Yet throughout this period, there has been remarkably little interest shown in the persuasive powers of the preacher's voice. Why? Because the Qur'an has always been deemed to be sublime and eloquent in itself. The divine message has never required rhetoric to be sold; what it has needed is the right sort of listening. So the kind of guidelines that we found devoted to speech-making in Athens and Rome are, in the Muslim world, devoted to the listener's role. If anyone remains unconvinced by the Qur'an, the fault will have been, not with the words, but with the organ of reception – the human heart – and its inability, or its refusal, to work properly.[17] Which is why, according to one source, such people 'have hearts wherewith they understand not, eyes wherewith they see not, and ears wherewith they hear not'.[18] Hearing, then, is not a passive activity; it's just as much an action as speaking. One Islamic writer likens listening to a link in an electrical circuit, in that no understanding can flow freely unless the circuit remains unbroken. There's nothing second rate about being part of an audience in this tradition.

Nor, it seems, is listening a poor relation to speaking if we move even further east, and examine early Buddhist traditions. It's not just that Buddhist ceremonies have always been rich in all sorts of sounds; there's also a deep veneration of listening. Just take a look at a typical statue of Buddha, or of a Buddhist sage, which you might come across in a temple or museum. This icon is instantly recognisable because the figure is invariably sitting cross-legged and looking straight ahead. But a Buddha usually has numerous other bodily marks: flat feet, deep blue eyes, large eyelashes, a certain number of teeth, and so on. Each of these marks is symbolically significant: it's a mark of destiny, which hints in one way or another at his greatness. But one bodily mark is especially striking, I think: the Buddha's extraordinarily large and long ears. These, it seems reasonable to assume, are the signs of a good listener. It's significant, for instance, that the most ancient Buddhist suttas all begin with the words 'Thus have I heard'.[19] We would have to conclude, I think, that this is a religion, like Islam, which has always placed great emphasis on listening. And, again, it's an emphasis on listening as an active thing, a difficult, skilful pathway to eternal truth, as its followers understand it.

It would be tempting to regard the Islamic and Buddhist veneration of listening as representing a complete contrast to what had happened back in Rome, or in Athens, where oratory seemed to reign supreme. Yet these weren't ever cities where speaking mattered and listening didn't. Plato had always favoured dialogue above all else: speech and listening as the two ingredients of reasoned discussion, where people present arguments and other people affirm or refute them by means of logic. Speech, then, was always supposed to be more than one-way trickery. It was about talking *to* one another, not *at* one another.[20] It was, in short, about a conversation. And I would go as far as to say that it was through conversation, as much as through writing, that we should locate the origin of philosophy and rational thought in the ancient world. The habit of talking together – and of questioning each other – would surely have driven forward knowledge, and, perhaps occasionally, mutual understanding.[21] Certainly, our current generation of classical scholars stress the importance of conversation, dialogue and listening as much as their predecessors would have stressed rhetoric. In the ancient world, says Simon Goldhill, it was dialogue that was at the heart of civic practical reasoning, and 'speech-evaluating' was as important a part of democracy as 'speech-making'.[22]

Did conversation die when Rome fell? Hardly. You could even say it has been revived in our own mass-media age. After all, at their best, radio interviews or TV chat shows are robust modern examples of the ancient art of dialogue. As for the speeches of politicians, even they can be thought of as the sweet sound of conversation rather than as one-way noise. When Obama repeats the phrase 'Yes we can', it echoes the call-and-response preaching of many American churches, especially African American churches. It's a conversation. Woven through his words and his voice is what one observer calls 'a respect for dialogue'.[23] It's precisely why, as Zadie Smith showed, he rarely speaks with just one voice, even if his language is English. Through him, we hear a multicultural nation speak.

8

Babble: The Noisy, Everyday World of Ancient Rome

In the centre of Rome, just to the west of the River Tiber, is the district of Trastevere. It's a place that bustles with life, especially after dark. The narrow streets and small, enclosed squares are packed with shops open late, and hundreds of bars or restaurants, each spilling chairs and tables out on to the cobblestone streets, and sucking in and spitting out a steady stream of diners and drinkers. Some of the people here are local; but many more are visitors who have flocked to Rome from all over the world. You will hear not just Italian, but French, English, Japanese, Spanish, Russian and Arabic voices – all blending to create a particularly exotic babble of human chatter. And, every now and then, the clatter and rumble of delivery vans, or street-cleaners, of shutters going up or down, doorbells ringing – all the backstage noises of a modern commercial district bringing in its supplies and having its detritus removed.

With its richly layered buzz of human interaction, machinery and money-making, the soundscape of Trastevere is in many ways the quintessential soundscape of the modern European city. But it's also not so very different from the sounds we would have heard here about 2,000 years ago. Ancient Rome wasn't the first city in human civilisation, of course. The *Epic of Gilgamesh* refers to the 'uproar' and 'clamour' of the city states of Ur, Uruk and Sumer 1,000 years or so earlier.[1] But, at the height of its power and wealth, Rome was in a league of its own: a city of one million people – the largest city known so far, and by all accounts the loudest.

> Many an invalid dies of insomnia here … The endless traffic in narrow twisting streets, and the swearing at stranded cattle … I'm forever trampled by mighty feet from every side, while a soldier's hobnailed boot pierces my toe … There's a hundred diners each followed by his portable kitchen … Meanwhile his household, oblivious, are scouring the dishes … are clattering the oily back-scrapers … The slave-boys bustle about on various tasks … The impudent drunk's annoyed if by chance there's no one at all to set upon, spending the whole night grieving, like Achilles for his friend, lying now on his face, and then, turning onto his back: since it's the only way he can tire himself; it takes a brawl or two to send him to sleep …[2]

The writer Juvenal was prone to exaggeration – and scornful of the imperial Rome in which he lived. Yet, he captures something of the energy of that overcrowded, restless city. In ancient Rome, noise was a sign of vitality. But when it was so constant, and difficult to get away from, it was also a source of irritation. It's therefore no wonder that when we read about the sounds of ancient Rome, a world of simmering social tension is revealed.

One of the most vivid descriptions of the Roman sound-scape comes from Seneca, who describes living directly above some public baths:

> Just imagine the whole range of voices that can irritate my ears. When the more muscular types are exercising and swinging about lead weights in their hands, and when they are straining themselves, or at least pretending to strain, I hear groans. And when they hold their breath for a while and let it out I hear hissing and very hoarse gasps. But when I have to put up with an unathletic fellow, one satisfied with a low-class rub-down, I hear the slap of a hand pummelling his shoulders … Imagine also a quarrelsome drunk, or sometimes a thief caught in the act, or a man who loves to sing in the bath. And then imagine people diving into the pool with a great splash of water. Besides these men whose voices are, if nothing else, at least natural, imagine the hair plucker with his shrill and high-pitched voice continually shrieking in order to be noticed; he's never quiet except when he's plucking armpits and forcing his customer to shriek instead of him. I could wear myself out just listening to the variety of shouts among people selling drinks, sausages, pastries; each restaurant or snack bar has its own hawker with its own recognized jingle.[3]

Driven to distraction by these noises from below, Seneca might have sought some respite by going out for a walk. If he did, however, he would have been hard put to find any in the streets around his apartment. Outside, there would have been even more people hawking their goods, each vendor with his or her own distinctive call – the louder the better in that constant struggle to attract customers. An area such as Trastevere provided an important trading post for the whole city: it was a place of loading and unloading, of workshops, a melting pot of cultures – and filled with people, among the poorest, who lived cheek by jowl. As long as the empire was growing and thriving, Rome sucked into its orbit more and more exotic goods, tastes, smells, colours and sounds – all of which drifted, trundled or roared their way along streets so narrow it was supposed to be possible for someone living upstairs on one side to reach out and touch their neighbour opposite.

At ground level, there would have been a stream of bellowing animals, herded from the countryside towards their slaughter, either for sustenance or sacrifice;[4] and, since this was an imperial capital, bringing economic migrants, traders and slaves from provinces all around the Mediterranean and beyond, people chatting everywhere in dozens of languages – Syriac, Coptic, Punic, Celtic, Hebrew. There was Latin being spoken, too, of course, but rarely the refined and orderly version we see chiselled on to great monuments or hear recited in today's classrooms. The graffiti found in brothels or barracks provides a snapshot of what was a living, earthy language. There were secret codes and different styles used among the various trades or professions or in different city districts; rules of syntax were casually discarded; free reign was given to verbal creativity. As Jerry Toner points out, everyday speech was 'full of necessity and urgency'. In the streets and workshops we would have heard lots of double negatives – 'never do no one a good turn' – and consonants dropped for ease of speech: *scriptus* becoming *scritus*, *sanctus* becoming *santus*, *hortus* becoming *ortus*.[5] New sounds were sometimes added for dramatic effect: why *fugio*, run away, 'when you can *fugito*, run like hell?', Toner observes.[6] And why just talk if you can also gesticulate noisily, complete with hand thumping, chest or forehead slapping, and foot stamping? In the streets of Rome, human communication was conducted with 'an emotional intensity and immediacy' that must have been hard to ignore.[7]

How on earth could people get on with their normal life surrounded by all this? According to Dio Chrysostom, the

Greek observer of Roman ways, it was by cultivating an air of obliviousness, even in the thick of turmoil:

> I remember once seeing, while walking through the Hippodrome, many people on one spot and each one doing something different: one playing the flute, another dancing, another doing a juggler's trick, another reading a poem aloud, another singing, and another telling some story or myth; and yet not a single one of them prevented anyone else from attending to his own business and doing the work that he had in hand.[8]

This roiling cauldron of people and sounds had a certain visceral appeal. Seneca, for instance, hints that he found at least some sensual pleasure, not in the grand architecture and clean, ordered, open-air Roman forum, but slumming it in the city's alleyways and bars and brothels, as well as in its steam-filled baths, where he could, as he put it, catch the smell of perfume on a passing stranger, the wine on his breath, the belch of a drunken man.[9] Even the most elevated were happy to add the noise of their own revelries to Rome's rich sonic brew. The emperor Augustus, for example, is thought to have enjoyed pantomime. And emperors set fashions. The city hadn't really ever embraced song and dance as the ancient Greeks had, but following Augustus' enthusiastic approval of the art form, audiences would join in with the singers and dancers on stage with such gusto that it was not uncommon for performances to end with a full-blown riot.[10] Such passionate embrace of sensory delight – even if that meant something risqué or raucous – was part of being Roman. Which is no doubt why Seneca, Stoic that he allegedly was, boasted that he 'took no more notice' of the roar of the gym or the street than he did of 'waves or falling water': to be distracted by noise, he reckoned, was to succumb to one's own inner disquiet.[11]

Enter the ruins of an old tenement building on the Via

Giulio Romana near the forum, however, and it is difficult not to think of Seneca's tolerance of noise as the slightly indulgent pose of someone with the privilege of being free to choose when, and when not, to slum it – someone with little awareness of the realities of plebeian life. It was once a five-storey block of flats: some of the floors would have been shops, others filled with cell-like rooms, others with slightly more spacious units, but all separated by only the narrowest of corridors – and right next to a street.[12] Those who had no choice but to live here nearly 2,000 years ago, in the cramped and tottering apartments, very close physically to the grandeur of the forum but socially a world apart, would never have enjoyed peace or quiet. And the experience of those unfortunate souls living in this particular spot would have been repeated across the city. One official census from later in the imperial period suggests that Rome had fewer than 2,000 single-family homes, but over 46,000 shared dwellings in apartment blocks.[13] They were poorly built, sometimes with paper-thin walls that would have provided little protection from the sounds of the street, the shops and workshops close by, or the endless delivery carts, which, some of the records seem to suggest, Julius Caesar had ordered should come into the city only at night. Bedrooms were non-existent: sleeping was done in shifts, men, women or children occupying rudimentary recesses in the walls of other rooms. These Romans, at least, lived with little privacy and no means of keeping at bay the noise of family members, let alone neighbours.

This close proximity must have been a breeding ground for irritation and intolerance as much as for Seneca-like stoicism. Predictably, scapegoats were sought, and those who seemed a little different were the most vulnerable. We catch a whiff of this in Juvenal. From his wealthy, privileged perspective, at least, he looked down at the Syrian community in Rome as a form of sewage washed in from the Mediterranean, with 'its lingo and manner, its flutes, its outlandish harps … its native tambourines, and the whores who hang out round the race course'.[14]

It's hard to know if Juvenal reflected the view of poorer people actually living among the Syrians. But it's not hard to imagine that in cramped conditions it was often every man for himself. Those with a little power or influence used it. The education lobby saw to it, for instance, that no coppersmith could set up shop in a street where a professor lived, so that he at least could study in peace, even if no one else could. Others just escaped the noise and the overcrowding – an early example, perhaps, of that troubling mix of prejudice and fear called 'white flight'. Seneca preached tolerance. But in the end even he decided to move from his apartment above the gym and settle somewhere a little more relaxing. 'Why should I need to suffer the torture any longer than I want to?' he explained, with what can only be described as a complete lack of stoicism.[15]

If you were among the wealthiest of Rome's citizens, it was often to places such as the Palatine Hill, overlooking the forum, some thirty-two metres below, that you would have turned for a refuge from the noise of the common people. Like the forum, the Palatine is now largely a maze of ruins. But if you visit it today, you get a real sense of the openness, the fresh air and relative calm that you would have experienced as you wandered among the most desirable of addresses. At least until the end of the first century BC, the grand houses of the Palatine Hill would have been oases of quiet, punctuated only by footsteps on marble, or the Romans' favourite sound of all, the trickle of ornamental water.

By the end of the first century AD, the private houses of the rich had given way largely to even grander imperial palaces. There are surviving traces of elegant terraces at a palace built by Tiberius. Nearby, at the Domus Augustana, the private residence of the emperor, there are outlines of a vast central courtyard, of private baths, fountains and colonnades. Inside the palace, or the private mansions they replaced, most of the rooms would have had large woven hangings to absorb sounds and create a little privacy. There would have been slaves and servants toing and froing, of course; but their presence would have been muted. Their masters and mistresses often insisted that the slaves attending them at meals be silent. This was a silence loud with meaning, for it boasted to any guest of the householders' healthy degree of control over their own domain. As Seneca observed, in these circumstances, 'A whip punishes any murmur and not even accidents – a cough, a sneeze, a hiccup – are let pass without a beating.' [16]

Secure in a lofty, tightly regulated zone of private sensual pleasure such as this, the rich, powerful and educated would amplify their social – as well as their geographical – distance from the plebs below. Being noisy was condemned as vulgar – as bad, really, as being smelly. Writers such as Ammianus portrayed the snorting, grunting and quarrelling sounds made by ordinary Romans at play as bestial, if not obscene.[17] Breaking wind was a complex matter. It was generally understood that one had to try hard to keep it all bottled up when in polite society. Yet there were rumours people had died from the strain involved; and there was always the risk that the noxious vapours trapped in one's body might go straight to the brain and disturb one's whole body and mind. Perhaps, Emperor Claudius wondered, even the best people should be free to fart. Generally, though, social distinctions were worth keeping up. And other people were simply too dirty, too rough, too loud, too close for comfort. Whenever the great and good of the Palatine had reason to pass through the thronging streets of the central district, they would employ retinues and retainers 'to doughnut them and so re-establish distance between themselves and the crowd'.[18] Predictably, pulling up the drawbridge did nothing to enhance mutual understanding or sympathy. It's possible that any ban on wheeled vehicles in the streets during daytime was designed merely to satisfy the plutocrats' desire to keep streets clear for their own horses; there was little regard for the thousands of ordinary Romans thus condemned to nights of broken sleep. But then again, how could someone whose sleeping quarters were so far away from the main thoroughfares begin to imagine the wearing effects on others of endless night-time noise?

It would be wrong, however, to conclude that the rulers of ancient Rome were indifferent to the cultural or political value of sound. They would have heard voices warning them of sensual chaos and hedonism spinning out of control, of the spread of demoralising and effeminate music destroying the virility on which an empire had been built. They would have believed that sounds weren't just external things to be noticed in a detached fashion: they invaded the body – and they were capable of corrupting – or, alternatively, of cultivating – minds. So it always made sense, as far as the Roman elite was concerned, to put noise to some sort of moral purpose. That meant, in their case, using it to maintain social order – using it as part of a wider plan to nudge people into taking their pleasures in a much more controlled way.

So the public centre of ancient Rome wasn't abandoned. Instead, it was transformed into a 'stunning stage set'.[19] A programme of beautification had begun under Pompey and Julius Caesar. But it was under Augustus and his successors that large numbers of statues were put up, gardens and baths opened, temples and theatres and long straight streets built – all for the public's pleasure. They formed the impressive backdrop for performances rich in organised sound. Music and dancing accompanied the unveiling of statues, helping to bring them to life.[20] Processions continued a long tradition of hauling the spoils of war through the city. These were crowd-stopping affairs. An account of one earlier procession, celebrating the defeat of the last king of Macedonia in 167 BC, described the endless succession of chariots, each one laden with captured arms and armour:

> ... helmets lying upon shields and breast-plates upon greaves
> ... shields ... horses' bridles ... swords ... long Macedonian
> spears ... all the arms being loosely packed that they smote
> against each other as they were borne along and produced
> a harsh and dreadful sound, and the sight of them, even
> though they were spoils of a conquered enemy, was not
> without its terrors.[21]

A little further back in the parade, there would be trumpeters; soldiers singing victory songs either in praise of their general or, more often, ribbing him; musicians; dancers; livestock to be sacrificed – all heard and watched by cheering crowds.

Such noisy displays were meant to be at least a little terrifying, as well as enjoyable. They reminded citizens of the power of Rome, the opulence that flowed from the wealth of nobility and the might of the army. They reveal to us that, even in the overcrowded chaos of the world's first metropolis, sound was never just sound. Wherever we find it, it was loaded with meaning, whether helping to measure the social gulf between rich and poor, or offering the tantalising hope, for those in charge, of creating an orderly, crowd-pleasing landscape of the senses. Of course, the biggest public events of all took place at the Circus Maximus and, a little later, at the famous Colosseum. So we should turn next to discover what happened when the rulers and the ruled came face to face: when the crowds roared with emotion amidst the violence, spectacle and noise of the Roman Games.

9

The Roaring Crowd

During the London 2012 Olympics and Paralympics, it was impossible not to notice how often athletes, TV commentators and spectators mentioned the sound of the crowd. In the main stadium, the Velodrome and the Aquatics Centre, competitors, especially those from the home nation, kept saying how they were taken aback by the sheer volume of the roar – lifted by it, pushed on to victory by an emotional wall of human noise. Occasionally, too, those from other countries would speak of being a little intimidated. Then there were those isolated but mesmerising moments when the crowd turned against someone and showed their displeasure vocally: the Chancellor of the Exchequer George Osborne being booed while he waited to hand out medals at the Olympic stadium, for example – perhaps for his policies, perhaps simply because he's a politician. In any case, he learned the hard way that even during the Paralympics, a stadium full of people gives voice to more than just sporting passion. In this cauldron of concentrated sound, all sorts of emotions get mixed up and run high. People speak their minds through noise, without necessarily having planned to do so.

It's something the imperial rulers of Rome would have understood 2,000 or so years ago, for it's back then that the biggest amphitheatre the world had known so far was built. And it's there, during the Roman Games, that we get our first visceral experience of the power of an audience – the power to become a collective force playing its own vital role in the volatile politics of the arena.

The Colosseum wasn't the first venue in the Roman world to stage gladiatorial combats and other games. Amphitheatres had been around for years, and not long after the Colosseum had been opened in AD 80 by the Emperor Titus, countless imitations sprang up all over the empire. But the Colosseum was the largest, the most famous, the most advanced, the most spectacular of them all – the model for all sporting stadiums to this day. Even now, when only part of the Colosseum's struc-ture survives – the central floor has gone – and the stands all around are dotted benignly with tourists, the building is capa-ble of conjuring up something of the electrifying atmosphere that must have taken hold of this place nearly 2,000 years ago when the biggest Games took place.

If you were a member of the crowd back then – heading through a numbered entrance, weaving along a series of painted corridors, up ramps to the right section and row, and then, finally, reaching your own numbered seat – you would have emerged into a stadium heaving with colour and pageantry. It was, as Keith Hopkins and Mary Beard describe, 'a brilliantly constructed and enclosed world, which packed emperor, elite and subjects together, like sardines in a tin'.[1] The spectators would have been arranged hierarchically by status, in steeply serried ranks: senators in their white togas with red borders; equestrians at the front; plebs higher up; and slaves, the poor and women crammed in at the top.[2] Thus arranged, some 50,000 people would have faced each other, eagerly awaiting the promise of a day-long visual feast of colourful costumes and gore. Like those gathered in London's Olympic stadium, they would also have been about to be enveloped in a barrage of sound – one both exhilarating and alarming, one in which they would play a full and strikingly political role.

The aural assault would probably have begun even before the spectators had reached their seats. Milling around near the entrances, they would have heard an array of ugly backstage noises racketing around in the hellhole hidden away beneath the arena: slaves, gladiators, tethered animals, chained prisoners awaiting their fate, and the shouts of the hundreds of animal-handlers, guards and technicians working behind the scenes. They would probably have heard the grinding of machinery, too. The Colosseum had an elaborate system of hoists and cages, so that animals could be delivered efficiently from the basement to the trapdoors above.[3] It was, however, when the audience was seated in the arena and the show had begun that they would experience the full panoply of sensory thrills: the smell of blood and sweat and perfume; the sight of death and the flash of weapons; the sound of trumpets and drums; the screams of victims; and, above all else, the roar of the crowd.[4] Not every show was a blockbuster, of course. And the timetable of events across the day could vary considerably. But the success of the Games held there would depend on the appearance of familiar favourites as well as the odd surprise, so proceedings would frequently start with a noisy procession of politicians, religious figures, images of the gods and emperors, sponsors, musicians, gladiators and animal-handlers. Once a fanfare of trumpeters had called all of the spectators to attention, there might have been a morning of wild animal fights, with an execution – or maybe a little burlesque interlude – at lunchtime. As the climax to the afternoon, there was, perhaps, the promise of gladiatorial combat – an attempt to bring into the heart of Rome some of the visceral excitement of the battlefield.

It would have been hard to recreate the real experience of the Roman soldier, of course. A first glance at the stone carvings of battle scenes that spiral around the Column of Marcus Aurelius still standing in the Piazza Colonna shows the army as it would have liked to be seen: advancing into the fray in ordered ranks, disciplined, calm, unified, and fighting hand-to-hand with courage. Look more carefully, however, and you will notice scenes of barbarism and chaos. There's rape, pillage, the beheading of enemies who had surrendered, the carrying off of body parts as trophies, all under the watchful, approving eye of their commander.[5] During the actual fighting on a battlefield, such was the melee that the ordinary Roman soldier – hot and tired from the weight of armour and equipment, tightly packed in a phalanx, the air around him thick with kicked-up dust and the overwhelming din of clashing arms and screams and yells – would have got no real sense of the course of battle.[6] From this messy, confusing, heaving, brutal turmoil, the Games picked out and then ritualised for the crowd's delight and political education a theatrical taster of war, and indulged the Roman predilection for seeing one-to-one close combat with shield and thrusting spear as the epitome of heroism.

But the clashes and the screams of gladiators were a special treat. More often, the spectators' ears would have been assailed by the roar of wild animals. These were brought into the arena as 'living war booty', symbols of the conquest of distant territories and Roman mastery of the natural world.[7] The Colosseum was officially opened by Emperor Titus with a festival of slaughter in which, according to one later Roman historian, some 9,000 wild animals died. Even if that figure has been exaggerated, it's clear that day in, day out, more animal than human blood was shed in this arena. Lions, leopards, tigers, bears and bulls all featured regularly, either to kill or be killed.[8] For the crowds, the most persistent soundtrack to their day out was the dying groans of these creatures.

Perhaps the most exotic victims, however, were elephants. These magnificent beasts had long been used in battle, particularly by Indian warlords. Prince Chandragupta, who had established the Mauryan Empire across northern India and Afghanistan towards the end of the third century BC, was said to have maintained 9,000 elephants for his own army. We know from accounts from this period that they would have been sent into battle in their hundreds, perhaps thousands, heavily armoured, covered in neck ropes and bells, and with up to seven men on their backs armed with hooks, quivers, slings and lances, sowing terror and causing devastation among enemy forces whenever they were goaded into charging. Often they were used as royal mounts, older elephants being especially prized for their supposed ability to endure the confusion and deafening din of battle.[9]

The Romans knew all about this ancient Indian tradition through having read the histories of Alexander the Great's earlier campaigns in the East. After defeating Carthage, they also had access to their own supply of elephants from Africa. Once brought to the capital, these enormous creatures were particularly useful when it came to public executions. Army deserters were thrown under an elephant's feet to be crushed to death. It was said that military discipline always improved a little after one of these gruesome displays.[10] This was entertainment with an educational purpose.

Animal shows, gladiator fights, comic interludes, processions: all these spectacular and noisy events were staged by a ruling elite with an eye to some sort of political impact on the expectant crowds who witnessed them. Vespasian had ordered the Colosseum to be built as part of a wider city redevelopment designed to wipe away the memory of his predecessor Nero and his notorious private pleasure palace the Golden House. The new amphitheatre was a fresh start, providing Roman citizens with a pleasure palace all of their own. It wasn't for the few; it was for the many – indeed, for the whole of Rome. Of course, as a gift it was also, as Keith Hopkins and Mary Beard point out, 'a brilliantly calculated political gesture': an ostentatious display of imperial power and generosity, a bribe for the people's continuing loyalty.[11] As indeed were most of the Games subsequently held at the Colosseum and venues like it across the empire. In principle, it might seem rather foolhardy, dangerous even, to allow so many people to gather in one place, so near to their rulers. But by the time the amphitheatre opened, the emperors felt secure enough to risk, even enjoy, meeting 50,000 of their subjects face to face and seeing them entertained. It showed a degree of confidence that would surely impress any crowd.

Yet being here as part of the audience didn't always mean a supine acceptance of what was on offer. We know from earlier descriptions of Games held at places such as the Circus Maximus that watching the main event was only one part of the attraction. In the stands, you could meet and chat, and enjoy free handouts of food and prizes. There's even a detailed if slightly mischievous account from Ovid of how you might have used the crowds and the noises at these other venues as a cover for flirtation and illicit caresses. When the horses are running, he writes, 'opportunity awaits':

> Sit as close as you like; no one will stop you at all … contact
> is part of the game … Try to find something in common,
> to open the conversation … when the gods come along in

the procession, ivory, golden, out-cheer every young man,
shouting for Venus, the queen ...[12]

It's easy to imagine the same sort of shenanigans going on at
the Colosseum. Yet, wherever the action was taking place, it
must have been hard at times not to be swept along by the
heightened emotions and adrenalin rush of the main show.
You would enjoy rooting for your favourites, and experiencing, vicariously but powerfully, the feeling of danger. But a
sense of collective power would also then take hold, the noise
you made together determining the fate of those performing
below. At the end of a gladiatorial combat, the presiding master of the games – the 'editor' – would probably take account
of the crowd's mood before deciding whether a defeated
fighter should be dispatched or not. A loud shout of 'Mis-
sum!' might save him; a shriek of 'Iugula!' might prompt the
fatal blow.[13] Yet nothing was ever entirely predictable. When
so many people were so tightly packed together, sheer proximity turned an audience into something much more volatile.
Moods could turn.

One rather horrifying example of events spinning out of
control took place not in the Colosseum but – possibly – in
the nearby Circus Maximus, an even bigger outdoor arena in
which more than 200,000 spectators might have gathered at
any one time. In 55 BC, Julius Caesar's powerful rival Pompey
put on a show for the Roman people that featured the slaughter of somewhere between seventeen and twenty elephants
– the sources differ on precise numbers. At first, apparently,
all was going well. The crowd seemed to enjoy the sight of an
elephant crawling on its knees, too wounded to stand up but
still just able to snatch shields from its opponents and throw
them into the air like a juggler. But then some elephants tried
to break out of the palisade that enclosed them, causing what
Pliny, with what must be a classic bit of understatement, called

'some trouble' in the crowd.[14] Even worse, having seen the elephants putting up such a brave fight, the crowd were unsettled when they had to witness their awful death throes. Pliny said they were simply 'indescribable',[15] but we have a pretty good idea from other sources that it wouldn't have been a calm or peaceful affair. Accounts of Indian battles recall elephants uttering cries 'like cranes', of shrieking aloud as they were pierced by swords and lances and arrows, of running amok.[16] George Orwell wrote a classic account of how in the 1930s, when posted as a colonial policeman in Burma, he had had to shoot a rogue male elephant and listen to its agonisingly slow, un-silent death, some thirty minutes after the first bullets had struck:

> It was obvious that the elephant would never rise again, but he was not dead. He was breathing very rhythmically with long rattling gasps, his great mound of a side painfully rising and falling …
>
> I waited for a long time for him to die, but his breathing did not weaken…

Orwell then tells of firing two more shots:

> His body did not even jerk when the shots hit him, the tortured breathing continued without a pause. He was dying, very slowly and in great agony, but in some world remote from me where not even a bullet could damage him further. I felt that I had got to put an end to that dreadful noise. It seemed dreadful to see the great beast lying there, powerless to move and yet powerless to die, and not even able to finish him. I sent back for my small rifle and poured shot after shot into his heart and down his throat. They seemed to make no impression. The tortured gasps continued as steadily as the ticking of a clock. In the end I could stand it no longer and went away.[17]

Spectators at Pompey's show in 55 BC wouldn't have been able to retreat quite so easily. And so, as Pliny records, the elephants, trumpeting, moaning and wailing as they crumpled, were able to play 'on the sympathy of the crowd'. According to Cicero, there was 'an impulse of compassion, a feeling that the beasts had something human about them'.[18] The result, Pliny tells us, was that the show backfired horribly: 'forgetting Pompey and his lavish display specially devised to honour them, [the spectators] rose in a body, in tears, and heaped dire curses on Pompey, the effects of which he soon suffered'.[19]

Of course, Pompey's fate wasn't really decided there and then, solely by the dying wails of an elephant. But the story shows how the wisest political leaders of the day could sense the will of the people if they kept their ears open at gatherings like these. And in the enclosed world of the Colosseum, the audience mood would have been even more intensely felt. The ebbs and flows of emotion would have rippled around the Roman arena minute by minute; not just the cheers and applause for the latest stars, but the uproarious laughter ridiculing unpopular figures, the hisses and catcalls. It would have been hard, no doubt, for those on the wrong end of the crowd not to feel hurt, maybe even vengeful, but you could bet that reacting badly would have made matters worse.[20]

So were the various noises of the audience at the Roman Games a tantalising pre-echo of the noise of people power in action? Not quite. After all, by the time the Colosseum was built, Roman autocracy was pretty firmly entrenched. It's true that the crowds made good use of the chance to express their feelings, letting their views be heard in the most visceral way. But even in this setting they could be manipulated. Whole blocks of seats could be allocated to friends and followers of the organisers. Stooges could be dotted about in the crowd, paid to start applauding – or booing – at all the right times. Soldiers could even be deployed in the stands, to strike down any member of the audience who flagged in their applause for the emperor. Whatever devious technique was used, a few rehearsed chants and a ripple or two of manufactured cheering were often enough to create a misleading impression of the will of the people.[21]

The Colosseum, then, provides a wonderful template for all those emotional roars of excitement, upset and disapproval we have heard in our own lifetimes, rippling across the venues of the London 2012 Olympics, or even through the TV studios hosting their talent contests. The Roman amphitheatre, though, was, as Hopkins and Beard point out, 'very much more than a sports venue', and very much more than a pleasure palace.[22] It was also, as has been indicated, a deeply political arena – not for the unsullied expression of ideas or debate, but for an important bit of political theatre. As they chanted or jeered, members of the crowd could fantasise that their opinions and feelings mattered, that they were expressing their collective muscle as the Roman people. An emperor could imagine that he, too, was successfully parading his own power before his subjects and citizens. There was truth – and delusion – in both camps, for the Colosseum amplified in the most dramatic way that no one fully owned or controlled the world of the senses. The roaring crowd was neither as free as it imagined, nor as obedient as the ruling elite would have wished. Sound was its own arena – and, in it, power was delicately poised. In listening to the Roman audience at the Games, we hear both how noise can be used to govern us, and how we, in turn, can use it to resist.

10

The Ecstatic Underground

There are parts of every town and city which are secret – or, if not secret, at least hidden from everyday view. By visiting them, we leave behind the public soundscape of streets and stadiums, and discover the subtler but no less important soundscape of people's private worlds – a place where personal beliefs are played out in the form of mysterious rituals.

In ancient Rome, the most famous and perhaps the most misunderstood places of secrecy are the catacombs. These are complex networks of underground passages and rooms built between the first and fifth centuries AD, which extend over many levels and for hundreds of kilometres just beyond the old city walls. Some can easily be seen by tourists; others are shut, or can be visited only by special permission of the Vatican.

The so-called Catacombs of Priscilla, located a few metres beneath the Via Salaria, are among the less commonly explored sections of these ancient passageways. Their layout is fairly typical: a warren of dark, damp and chilly passageways, lined on both sides with graves in layered niches, and opening occasionally into larger rooms, each of which is stuffed full with inscriptions and frescoes, just visible in the dim light. In one of these rooms there is a scene of Moses striking water from a rock; on an archway, a painting of what looks like the Madonna and child with the Magi; on another wall nearby you can find Daniel among the lions; the resurrection of Lazarus; the sacrifice of Isaac; Noah, and even what looks like the breaking of bread during the Eucharist.[1] These images tell us that we are in the world of the first Christians of Rome – the followers of a new mystery cult who were for many years persecuted for their beliefs; people, in other words, who had to keep a low profile.

It's tempting to imagine these catacombs as Christian hiding places, where illicit rituals could be performed away from prying pagan eyes and ears. However, there is no evidence of elaborate ceremonies being held here, or of Christians seeking refuge underground. The catacombs are simply resting places for their dead – hundreds of thousands of them, stacked plainly in tiered niches or sometimes in more lavish family vaults. The real significance of this underground city lies in what it tells us about early Christian life above ground in Rome – and in the hints it gives us as to what it would have sounded like, and how important sound was in all its rituals.

Ironically, our clues are visual. Take a closer look at the walls, and suddenly it's not clear whether what we are seeing is the painting of the Eucharist or a pagan banquet at a wedding or funeral. The inscriptions are not all in the Latin of Rome; many are in Greek. Some burial chambers are Jewish in design, not Christian. In fact, much of the imagery down there draws on traditions stretching back in time to the first millennium

BC and from as far away geographically as North Africa and the Middle East. Rome was the cosmopolitan centre of a vast empire, and its people had a rich folk culture in which different religious traditions, in particular, were constantly being mixed up. So the sound world of the first Christians was unlikely to be filled with the subdued voices, measured singing and solemn prayers that would later echo through the medieval churches and cathedrals of Western Europe, and which we still imagine today as being the religion's essential sound. It was more likely to be wilder, more eastern in flavour, more pagan. It wasn't yet clear which religions – and which religious soundscapes – would survive, and which would not. In ancient Rome, Christians were still finding their way. Although they were starting to define their own rituals, everything was still fluid. And, as with all cults in the metropolis, they were shopping around for inspiration, a phenomenon we would recognise throughout history.

Walk into any modern-day spiritual or new-age bookshop in London or New York, for example, and you will be faced with an eye-watering choice of alternative beliefs and therapies for sale: notices advertising healing sessions that draw on ancient Chinese spiritual practices, perhaps, or therapeutic touch workshops inspired by Indian traditions; classes on Buddhist meditation, chanting sessions, introductions to shamanism and insights into reincarnation. Alongside an array of books covering every conceivable religion, you might also have the chance to buy incense, statues, oils, singing bowls or even a make-your-own altar kit.

Two things are really striking about all this. First of all, no one will tell you what to put in your shopping basket: there's a free-for-all, pick-and-mix approach. Secondly, most of what's on offer is looking to the ancient past as a way of establishing its credibility. A lot of what's available also looks very specifically to an eastern past – to traditions of Vedic healing in ancient India, to the spiritual powers of the Hebrews, the Persians, the Chinese, and, especially, to the hypnotic qualities of primordial sound.[2] And I can't help thinking that what we find in London or New York in the twenty-first century, say, provides a modern vision of the promiscuous marketplace of ideas and beliefs that characterised the most culturally diverse metropolis in the world some 2,000 years ago.

We can start to unravel some of the complexities of religious life among Rome's early Christians if we take a look at the church of San Clemente, one of the oldest churches in the whole city. Again, to understand what happened on the surface, we need to head downwards, if only because the accumulation of flood debris and human detritus means the city has risen some nine to fifteen metres over the past 2,000 years. So although at ground level the San Clemente dates from the early twelfth century, if you descend a staircase near the sacristy, you will discover that it's built on top of an earlier fifth-century basilica. Descend from another staircase at the far end of the aisle and you will reach even older Roman remains, from the first century AD, including those of a large villa which must have been the private house of a wealthy Roman family.

In what was one of the villa's central rooms there are some distinctly *un*-Christian signs: an altar featuring the slaying of a bull, torchbearers and a large snake; what look like benches for a large dining table; and a general decor and atmosphere that is rather cave-like. The altar, in fact, is of Mithras, and the room was evidently converted some time around AD 200 into a 'mithraeum'.[3] It's where followers of Mithras would have gathered for their rituals – which included sharing a meal at a table draped with the hide of a newly slain bull – at least until their cult fell out of fashion.[4] Although we don't know very much else about this villa, the fact that the church of San Clemente was founded right on top of it means it's likely that early Christians were also meeting somewhere inside.

Until their religion was officially recognized in AD 313, and they could build places of public worship confident that they wouldn't be persecuted or treated warily, Christians had to meet surreptitiously in so-called 'house-churches'. And this, it seems, was one of them.[5]

Why, one wonders, might there have been two different kinds of worship apparently taking place right next to each other – one involving Christians, the other followers of Mithras? Perhaps the two groups didn't quite coincide. Maybe new people with different beliefs moved into the villa, or the existing family converted, or various members of the household followed different cults at the same time. Whatever the truth, the mithraeum's presence reminds us that Christianity was just one of several cults springing up in Rome, living side by side, competing for attention. This profusion was partly the result of a relative lack of central organisation – at least to begin with. In these early days, each Christian house-church would have been somewhat isolated, operating almost like a guerrilla cell with only limited contact with any outside leaders. It was easy, then, for them to evolve differently, to take an independent line when it came to interpreting how exactly they were to behave. By the second century AD there were over twenty varieties of Christianity in existence in the empire.[6] That probably meant at least twenty different sets of rituals, and at least twenty different soundscapes of belief.

We would have to assume that, even if only for survival's sake, many of these nightly house-gatherings, or 'vigils', were usually pretty quiet affairs. And no doubt many early Christians were rather ascetic souls, spurning the Roman taste for the sensual, retreating into isolation and peacefulness, like St Theodore of Sykeon, who spent two years alone in a cave, or the first monks, who rejoiced in threadbare clothing and mute contemplation.[7] But the house-meetings of ordinary Christians – even the most sedate among them – could hardly have been silent. These were people who lived and worked in a society that still communicated largely by speaking aloud rather than by writing. The house-churches have therefore been seen as a bit like philosophical schools, where new ideas – or perhaps old ones from Egypt – about the afterlife might be debated vigorously into the small hours.

Given the early Christian expectation that Christ's return was imminent, there would have been lots of praying out loud, too. For all Romans – Christian or pagan – sound was an important means of communicating with and experiencing the gods and the supernatural realm. When Vesuvius erupted in AD 72, Dio Cassius tells us, people nearby assumed that they were hearing the din of war among the gods.[8] In every-day life, the city air was full of incantations, spells and curses; utterances were invariably needed to invoke the spirits. 'Wrap a naked boy in linen from head to toe,' went one Roman spell. Then clap your hands, and after 'making a ringing noise, place the boy opposite the sun, and standing behind him say the magic formula'. As Jerry Toner points out, magical and pagan practices had established sound as a vital if routine part of their rituals; it was therefore too deeply embedded a way of life to be ignored by Christians.[9]

Inside the house-church, then, muted chatter and prayer would probably have been punctuated by exclamations and chants, and of course the subtle sounds generated by the bod-ily movements and actions involved in important rituals. The Jewish use of water and oil, and bread and wine, continued; gradually added to this ancient mix was the use of salt, fire, candles, incense and ashes – ingredients that had also formed part of the sensory experience of pagan sacrifices. The regular celebration of the Last Supper evolved from the Passover into the Eucharist; the ancient practice of using oil to anoint some-one to a priestly or royal position was adopted for Christian ordinations or for people who had just died.[10] Christians, like all followers of a new cult, wanted to create brand new rituals in order to establish their own distinctive identity and appeal. Yet they also knew that in the struggle to attract followers, it made sense to adopt traditions that were comfortingly familiar.

In the cosmopolitan centre of a large empire, though, it was likely that in some Christian cells, these traditions included altogether wilder rituals that had come from outside the city – from places such as Greece and further east. There's evidence, for example, that many of Rome's early Christians drew on deeply held ideas of reaching a state of spiritual consciousness through 'ecstatic' behaviour: singing, dancing, weeping, wailing, speaking in tongues – the sort of unbuttoned behaviour the first leaders of the Christian Church would soon be trying to suppress as being distinctly un-Christian.

The simple truth is that, whatever later doctrine claimed, the religions which criss-crossed the ancient world had much in common. Most, for instance, made a great deal of the power of music and of vibration. In ancient Hindu traditions, as we've already discussed in Chapter 6, the power of a mantra lay not just in its words, but in the sounds of the words, the rhythms in which they were uttered. It was these elements, after all, which released vibrations into the world – and it was these vibrations that produced further, and tangible, effects.[11]

The whole idea of rhythmic sound unleashing spiritual forces would have made perfect sense to many pagans in Rome, who also believed that sounds, like smells, could have a direct impact – not just on the gods, but also, as they entered the body through openings such as the mouth and the ears, on mere mortals. If these sounds in some way represented dangerous or malevolent spirits, as they inevitably would from time to time, it was assumed they would need to be countered with other noises capable of keeping them at bay. Bells did this particular job very nicely. And indeed, as Percival Price has shown, 'nearly every society which has existed since Neolithic times has made and used bells'.[12] The earliest evidence for their use comes from China, where extraordinary powers were ascribed to them. Their sound was seen as 'a manifestation of universal essence'; ringing them helped sustain 'Universal Harmony'.[13] Hung under the eaves of temples or near sacred relics, they would also be used to drive away evil spirits.

This is, perhaps, why we find the same principle at work in the catacombs of Rome. Although, as I mentioned, these underground passages weren't used for elaborate rituals, Christians do appear to have rung bells or small tintinnabula as part of the funeral rites there from the very beginning. In this case, historians reckon that, just as in China, Christians believed – as their ancestors would have done – that 'the force of sound from the bell was used to keep demons away'.[14] Like much of what we now think of as distinctively Christian, the tinkling chimes of the tintinnabula were sounds that had already been heard by people all across the globe for centuries.

These chimes were just one end of the spectrum of noise. After all, if it was true – as common sense at the time suggested – that sound was a way of repelling troubling spirits, or attracting the attention of the gods, then what better way to guarantee access to the holy – including the Christian God – than by making a bit of a racket? Indeed, what better way to turn one's mind and body away from this ordinary world of work and routine and achieve some communion with the invisible world beyond than to fall into some kind of ecstasy, just like the followers of the Greek god Dionysus had done centuries before? After all, many of these Christians would have seen ecstatic release as embodying their own message of abandoning worldly cares or possessions and individual pride.[15] True, there probably wasn't much wailing and moaning during most Christian burials: given their belief in the afterlife, a wake should normally have been a time of remembrance and joy marking another step in the soul's journey rather than an end point to be lamented vociferously. But it was impossible to suppress completely the old pagan preference for wild mourning at the deceased having been snatched by Fate.[16] And it's easy enough to imagine other times when, during a house-gathering, the sharing of food and wine tipped into something a bit more sensual and chaotic – if not quite bacchanalian revelry, at least a little singing and dancing and letting go.[17]

We're frustratingly short of good eyewitness accounts to help us lift the veil on all these activities. As Barbara Ehrenreich notes, 'Most of what Christians of the first and second centuries actually did together … is unknown to us today.'[18] Yet the sheer regularity with which leaders of the Church later condemned wanton behaviour suggests that quite a few services had indeed been noisy and unrestrained enough to worry them. In the fourth century, for example, the Bishop of Caesarea, Basileios, got rather hot under the collar about Christian women dancing:

> Casting aside the yoke of service under Christ and the veil of
> virtue from their heads, despising God and His Angels, they
> shamelessly attract the attention of every man. With unkempt
> hair, clothed in bodices and hopping about, they dance with
> lustful eyes and loud laughter; as if seized by a kind of frenzy
> they excite the lust of the youths. They execute ring-dances
> in the churches of the Martyrs and at their graves … With
> harlots' songs they pollute the air and sully the degraded
> earth with their feet in shameful postures.[19]

As Ehrenreich points out, 'Whether the women's dances were really lewd or only appeared so' to Basileios is impossible to judge.[20] One suspects the latter. Yet the Church authorities were certainly right to see all around them the stubborn survival of pagan and mystical customs: not just Christians dancing wildly in circles, but Christians making offerings to shrines and vows to trees, praying to fountains, speaking in tongues.[21] Apart from anything else, this kind of behaviour allowed ordinary believers to think they were getting direct access to the deity. This left little room for a special caste of priests to stand above them as their guides and intermediaries, so it's no surprise to find that when Church leaders start appearing in the records – men such as the fourth-century Gregory of Nazianzus – we find them busily trying to enforce solemn kinds of behaviour:

> Let us sing hymns instead of striking drums, have psalms
> instead of frivolous music and song … modesty instead
> of laughter, wise contemplation instead of intoxication,
> seriousness instead of delirium … if you wish to dance in
> devotion at this happy ceremony and festival, then dance, but
> not the shameless dance of the daughter of Herod.[22]

As we know, music and song had been part of religious rituals for thousands of years. But, over time, as the different religions became more clearly defined, more regulated, more hierarchical, their attitude to the senses became ever more suspicious. In only a few hundred years, Muslim scholars would be struggling with the same dilemma confronting early Christian leaders. They would acknowledge that sounds such as music appealed directly to the listener at an emotional level. Was listening to it therefore dangerous, because it aroused unruly passions and distracted the listener from thoughts of God? Or was it – as the Sufi tradition within Islam believed – a means of moving the heart closer to God and to greater piety? That neither Muslims nor Christians could ever quite agree among themselves on the value of music's emotional force was one of the factors which, over the centuries to come, would create dangerous splits.

Yet the overall direction of travel was clear, even while those first Christians were gathering in that Roman villa now buried by the much grander church named after one of the first popes, St Clement. Order, routine, an official line on appropriate styles of speech and prayer and on what sounds could not be made: none of this had to wait until the world of literacy and the book. The oral society of the ancient world was perfectly capable of creating rituals and events that, over time, fixed beliefs and practices into quite rigid forms – indeed, of creating a world where some people had a greater right to make noise than others, and where listening was something that had to be done properly, according to rules and regulations.[23]

The soundscape of the early Christians of Rome captured that tension at the heart of all human-made sound throughout history. Sound travels freely through the air, disrespecting all physical or social boundaries, incapable of being easily contained. It is the simplest and most effective means of one culture learning about another, simply because it allows one culture to hear another, to experience it directly by letting it enter the body and mind. Yet it was precisely this promiscuous, disruptive quality that has always made it seem so dangerous to those seeking power. It's what has driven them to more and more inventive means of controlling it – even of trying, where possible, to seize for themselves a monopoly in noise. And it's this ongoing struggle – between noise as a spontaneous expression of the rough and tumble of human life, and noise as a tool of power wielded by princes and priests – that we'll explore in Part III, when we enter the world of sound in the Middle Ages.

III
Sounds of the Spirit and of Satan

11

The Bells

I f you had lived in the Middle Ages, one of the loudest
noises you would have heard – apart perhaps from thunder and earthquakes, or the awful din of battle – would
probably have been the sound sent flying out every day from
churches, temples and monasteries. Even if you only heard
it in the distance, it would have been an ever-present part of
your soundscape. For priests and monks – whether Taoists or
Buddhists in China, or Christians in Europe – the best way of
communicating with each other and to the people who lived
nearby was the bell. Each time one rang from the top of a
tower and its sound floated out across a village or town, religion's extraordinary hold over the secular world was signalled
loud and clear.

Yet the Middle Ages can't be reduced to a single soundscape, religious or otherwise. Churches always had to rub along with local customs and ancient traditions, and each town and village rang out differently. In parts of the Netherlands, throngs of pitched bells were popular, giving towns a musical quality. Elsewhere, according to various travellers, bells gave their towns a 'threatening' or 'plaintive' tone, or conjured up a distinctive air of urgency or calm.[1]

Nor was it always the clear, resonant chimes of cast-iron or copper that filled the air. In much of Eastern Europe and the Near East, for instance, the sound of Orthodox Christianity was not the pealing bell but the pounding beats of a hammer striking wood: the semantron. This is a wooden board a metre or two long, usually suspended horizontally from chains or ropes. When struck in different places, the semantron produces different tones, and with a skilled operator complex messages can be beaten out. It's now a much-loved part of the heritage of the eastern Orthodox religion, especially in Romania and parts of Greece; however, in the Middle Ages the semantron became popular almost by accident. This plain wooden object was a pragmatic response to metal bells being outlawed. Nowadays, the idea of banning something as harmless as a bell strikes us as bizarre. Back then, however, their banning can be seen as a measure of their power. They were loved, venerated and feared. They helped organise daily life, they were right at the heart of the struggle between good and evil, and they always had to be in the right hands.

This struggle for control over bells can be seen particularly clearly in Istanbul's fascinating history. One of the reasons this city is so richly layered with sounds today is that it's long been the meeting place of two continents – Asia and Europe – and the meeting place of two great religions – Islam and Christianity. In the Middle Ages, when it was still called Constantinople, the city was the centre of Orthodox Christianity – a place where some of the more ancient rituals since abandoned by the Roman church of Western Europe were proudly maintained – yet in the fifteenth century it came under the control of the Ottoman Turks. The great cathedral, the Hagia Sophia itself, was converted into a place of Islamic prayer, and across the city, mosques soon outnumbered churches.

There was an immediate impact on the soundscape of the city. Bells had been rung in most Christian churches here since they had first formed under the Roman Empire. But an Islamic council had already decreed in AD 630 that calling the faithful to prayer should be done only by means of the human voice.[2] This ruling had allowed a rich array of vocal tones and styles to flourish across the Muslim world, but it clearly had implications for Christians. It was all very well for them to ring bells quietly inside their own buildings, but outdoors the noise was anathema. So the biggest copper bells of the churches were silenced, and the wooden semantron – which had been around for centuries and was cheap to make – now came into its own, summoning followers without the piercing effect of a bell. Wherever the Orthodox Christian Church took root, so did the semantron, and as it spread, it adapted to local conditions. In Ethiopia, Christians would strike not wood but a special stone called the *dewall*. In Syria, Greece and Russia, they would sound bars, plates and rings of metal. When several were struck at once, complex melodies could be created. In Russia, where the monasteries were fiercely independent, a stunning variety of different rhythms, tone colours and pitches developed over time.[3]

The semantron, then, took the place of the bell in an important part of the medieval world. But to the east and the west of the Ottoman Empire, where copper or cast-iron bells were never silenced, we find a similar story: people putting their ringing to a rich variety of uses – and hearing in their sounds a whole range of subtle meanings that have now been lost.

For those who served God directly in the Middle Ages – the monks and nuns and priests themselves – bells were a vital tool of the trade. Monasteries projected an image of quiet seclusion, but for those who lived inside, often forbidden to speak, the routines of eating, sleeping and praying would have been regulated day in, day out, by the ringing of bells.[4] In Western Europe, Benedictine houses, for example, were amazingly precise in their practices. Guidelines, formulated originally by Christ Church in Canterbury in the late eleventh century, called for Benedictine monks to be stirred from their slumbers to sing prime, their first devotion of the day, by a small bell called the *parvulum signum*, rung quietly by the warden. Other small bells, the *skilla* or the *signum minimum* or the *minus signum*, and so on, would then ring out through corridors and around the cloisters, each one slightly different in tone and each one summoning the monks to one devotion or another at three-hour intervals. The end of daylight would be marked with vespers. Stretching into the darkness there would be compline at nine, nocturns at midnight, then, finally, matins at dawn – before the *parvulum signum* was heard once more and the whole cycle repeated.

If monks needed to be hurried along at any stage, they would hear the light-toned sound of the *tintinnabulum*, the direct descendant of the small Roman handbell. At the other end of the scale there was the large 'signal bell', which had a much deeper tone. This would have rung out across the whole monastery to summon everybody to a general assembly. There was also the *tabula*, a piece of wood struck with a mallet by the prior or abbot to announce the precise moment when monks were allowed to speak to one another in the cloisters. Eating together initiated a whole new ritual of bells: a gong to announce the meal, two strikes of the *tabula* to enter the refectory, a third strike when the meal was over and food needed to be cleared away, and finally a good long shake of the *skilla* to get everyone to say a last prayer before leaving. In other words, monks were told by bells when to wake, wash, say their prayers, come to meals, rise from the table, work and talk; they were moved around and kept in time by sound.[5]

Ordinary lay people would have got a small taste of this extraordinary life of bells whenever they went to their local parish church. They would be guided through the intricacies of each service, not only by the priest, but also with a series of auditory cues. A bell would sound to alert churchgoers to a particularly sacred moment, and they would know exactly when to bow or look up. The priest would also have tiny bells attached to his robes, tinkling as he moved about near the altar and pulpit. The richest array of sounds, however, would have been heard at the services held in Coptic and Orthodox churches in places such as Syria, Armenia and Georgia, where, together with cymbals, singing and intoning, bells created one of the world's richest soundscapes of devotion.

If the parish churches of the West couldn't quite compete with such a sensory feast for those gathered inside, they did at least have at their disposal the most powerful use of all for the very biggest bells in their possession: the summons to worship. One of the oldest parish bells in England can be found inside the church of St Lawrence in Caversfield, near Oxford. Cast in the early years of the thirteenth century, it's now standing at rest in a corner of the ground floor, but nearly 800 years ago it would have been positioned more strategically, near the top of St Lawrence's square Saxon tower. When St Lawrence's bell was in place and being rung, its rich treble tones would have been heard not just by those inside the church below, but by anyone outside for miles around: by those resting inside their homes, by those passing through the village on foot or horseback, by those toiling away in the fields nearby. Everyone would have heard it, ringing at noon or in the afternoon on workdays, in the morning on Sundays and holy days – and they would have known immediately to stop everything and celebrate Mass.

Bells had been used like this for thousands of years. It was common in ancient China to announce periods of worship through striking a bell. Christians seem to have adopted the practice first in North Africa during the sixth century: it's then that we have a priest in Carthage describing the 'holy custom' of local monks ringing a 'sonorous bell'.[6] It was indeed in the monasteries, rather than the parish churches, that bell towers were first built so that the sound of bells could travel as far and wide as possible. By the time village churches like the one at Caversfield, along with the abbeys and cathedrals, were all ringing from their towers – each competing for attention and prestige by increasing the volume and complexity of their peals, each offering more and more religious services to their parishioners – the air over the house tops in some medieval European towns and villages must have felt close to saturation.

Whether this symphony of sound irritated or pleased, it certainly served its purpose. The bell was a potent means by which a parish church, temple or monastery could project its power, define its territory and regulate behaviour across a whole neighbourhood. It served, for instance, as an official timekeeper to the community at large. In each monastery a monk would be charged with keeping vigil over a sundial or hourglass or candle and observing the position of the sun or the night sky, ready to reach for a bell rope to sound the hours at the right moment.[7] This lonely figure, marooned for much of the day and night high in a tower, was the heroic human predecessor of the mechanical clock.

Civil authorities also made use of the church's bells to mark time. In London, under William the Conqueror, the nightly curfew was rung out from St Martin's-le-Grand at 8 p.m., and then taken up by other churches nearby as a signal for all city gates to be closed. The pattern was repeated elsewhere, with minor variations. The curfew that fell across old Beijing in the days of the Mongol dynasty, for instance, was described by Marco Polo:

> There is a great bell suspended in a lofty building which is
> sounded every night, and after the third stroke no person
> dares to be found in the streets, unless upon some urgent
> occasion … Guards, in parties of thirty or forty, continually
> patrol the streets during the night, and make diligent search
> for persons who may be from their homes at an unseasonable
> hour, that is, after the third stroke of the great bell.[8]

In the West, too, being outside after the curfew sounded
wasn't just dangerous, it was deeply suspicious. By now, the
bell was saying, all decent people are indoors. And when a bell
was rung unexpectedly, this really caught everyone's attention: it signalled danger, or a death or perhaps a miracle having taken place.[9] Different messages often involved different
bells or different rhythms. People usually knew whether it was
time to get down on their knees and pray, to mourn, to panic
or simply to return home. They also knew the geographical
boundaries of their community. To be a parishioner was, in
effect, to be within earshot of the local church bell.[10]

By creating this bond, bells ensured the whole cycle of daily
life was framed, time and time again, in religious terms.[11] Yet
people also identified with bells because of something more
ancient and pagan in spirit, more deeply embedded in folklore:
they were in awe of their sacred power to dispel evil.

If you visit St Lawrence's church in Caversfield and look
closely at the 800-year-old bell there, you will notice a faint
inscription. It reads: 'In honour of God and St Lawrence,
Hugh Gargate and Sibilla his wife had these bells erected'.
It's a simple and modest dedication. But bells often had more
forbidding messages carved into them:

I disperse the winds

I put the cloud to flight

I break the thunder

I torment the demons

I put the plague to flight

My voice is the slayer of demons

Through the sign of the Cross let all evil flee.[12]

If these sound like spells, that is because they are. When a bell rang out, it was commonly believed that any words written on it would also be sent flying through the air to do their work. It's why in the fifteenth and sixteenth centuries the city authorities in Peking maintained an enormous bell weighing some fifty tons and covered both inside and out with lengthy quotations from Buddhist scripture.[13] It embodied the ancient belief that metal breaks magic and that noise drives away evil. Across China, for centuries, other smaller bells were sounded in ceremonies designed to control the weather, encourage good crops and create an aura around homes, sacred relics and temples, while Taoist priests wandered around with handbells. In the West, peasants who might once have referred to pagan 'spirits' talked now, under Christian influence, of 'demons' or 'devils'. But it was always bells that were trusted to keep these invisible forces at bay. The old Roman rite of purification of the crops every spring, for example, was simply converted and from the fourth century it became Rogation. In places such as Caversfield, priest and parishioners would have gone out from the church and processed into the fields, ringing their handbells to invoke blessings on the crops.[14]

Handbells, of course, had one great feature: they made the beneficial effects of a church bell wonderfully transportable, which is why they were among the most important possessions of itinerant Christian preachers and missionaries as they walked along country lanes from village to village. One of the most famous, still preserved in the National Museum of Ireland, is the Bell of St Patrick's Will. In Patrick's own time, it was probably just two bent iron plates riveted together, but it was thought capable of keeping away evil spirits, even curing illness. After many years this simple *clagan* or *clocca*, as the Irish called it, was considered so holy that it needed to be shielded from the gaze of mere mortals. It was subsequently coated in copper, and people were even appointed to be its keeper – a role considered so honourable that it was passed down from generation to generation.[15]

The sacred powers of even the humblest bells obviously made them objects of great reverence. But for the Church this created as many problems as it solved. For if bell-ringing worked for a priest, why wouldn't it work in the hands of, say, a simple peasant farmer wanting to protect his herd of pigs? The Church found an answer. It distinguished between bells that were blessed – which lay people could use for ordinary workaday miracles – and those that were baptised – which consequently had much greater powers but would remain firmly in the hands of the priesthood. Naturally, the Church ensured blessed bells were available for purchase at all good shrines. After all, there seemed no harm in making a little money from people's anxieties.

Anxiety is a crucial part of this story. The medieval world was steeped in superstition, and violent death always seemed just a hair's breadth away. In truth, the Church's social power drew strength from this state of anxiety. Every time a bell's power to dispel evil was proclaimed or performed, it would have reinforced the idea that demons really were all around, and that only the Church and its bells stood in their way. This was a message that would have brought people ever deeper into the Church's comforting embrace. It's one of the reasons why, in towns across Europe, houses were built close together around a church: people wanted desperately to be in the protective aura of the sound of its bells. As Thomas Aquinas declared, 'The atmosphere is a battlefield between angels and devils':

The aspiring steeples around which cluster the low dwellings
of men are to be likened, when the bells in them are ringing,
to the hen spreading its protective wings over its chickens:
for the tones of the consecrated metal repel the demons and
arrest storms and lightning.[16]

The same raging struggle between good and evil was played
out in sound right up to and beyond the grave. When someone was sick and needed communion, a bell-ringer would walk
before the priest as he came to visit. As death arrived, a 'passing bell', or 'death knell', would be rung to drive away those
evil spirits waiting to seize a departing soul; a different rhythm
or pitch would tell those nearby whether it was a man, woman
or child who had died. Other bells sounded during the funeral
procession and burial. Handbells would be rung around the
body to sustain the protective aura every time it moved. And if
the grieving family was rich enough, they would even pay the
church to have bells rung in perpetuity to protect the soul in
the afterlife.[17] When plague struck a community, the ringing
would sometimes continue until the bells cracked.[18]

In the Middle Ages, listening to bells reveals a world dominated by established religion, yet also suffused with ancient
superstitions. Underlying everything, however, was the feeling
that sound, moving invisibly through the air all around us,
connected people with each other and with objects. It was a
means of 'touching' at a distance. Sound enabled not just the
passing of information from one person to another, but the
passing on of other, less tangible qualities – especially, perhaps,
goodness or evil. As we will discover, it was a fundamental
belief that would shape many other aspects of medieval life,
from healing and entertainment, to architecture, to our relationship with the cosmos.

12

Tuning the Body

One of the fourteenth century's international best-
sellers was a book of anecdotes called the *Gesta
Romanorum* – the 'Deeds of the Romans' – and
among the stories it gathered up for its readers' delight was a
cautionary tale about the seductive power of sound. It tells of
the Emperor Theodosius, who was out hunting when he suddenly heard the sweet music of a harp and promptly rode off
with his horse to track it down. Theodosius reached a stream:

> When he arrived there, he perceived a certain poor man
> seated on the ground, having a harp in his hand. From
> hence arose a melody; and the emperor was refreshed and
> exhilarated by the delicious tones he created.[1]

The harpist turned out to be a fisherman, and he was unhappy.

For thirty years, thanks to God, the sweet chords of his harp had enchanted the fishes straight into his hand, so that he had been able to feed his family. But, he explained to the emperor, no more:

> … a certain whistler has arrived within these few days from another country; and he whistles so admirably, that the fishes forsake me and go over to him.[2]

The emperor gave the poor man a golden hook. The strumming of the harp worked its magic once more, and the whistler retired in defeat. The anonymous author of this tale then helpfully explains to his readers that they, poor sinners, were the fish, the emperor was Christ, the fisherman a priest, the harp was the word of God, and the whistler was the devil himself.[3]

The Middle Ages were awash with moralising tales, but what is fascinating about this story of the harpist and the whistler is that it shows a battle for the human soul where the chosen weapon for both good and evil was sound. The fish might be lured into damnation by the whistling, or reach grace by following the sweet music of God's words. It was all on a knife edge. In the fourteenth century anyone who read the tale would have understood that the noises you made – and the sounds that you heard – could send you either to hell or to heaven. Talking, playing music, listening: all these had to be done properly if you were to get along – in this world or the next.

Not many would have read the *Gesta Romanorum* for themselves, of course, not least because the book was in Latin. Its real audience was the priesthood. For them it was a training manual, and they could apply it in their daily work, for instance in the intimate space of the confession box. Perhaps it's fitting that it was in here – where your face was hidden and the priest could only hear your voice – that you were likely to be harangued at length about the temptations of the senses. You might be told, for a start, about the horrible sins of the tongue: idle words, boasting, flattery, lying, backbiting, gossiping, blaspheming. Then about the sins of the ear: eavesdropping, listening to the siren pleasures of music or the seductions of devilish words. Your priest might try to remind you of how Samson, Solomon and David had been overcome through the power of women's speech. You might be told that the ears were the gateway to the body, and through the body to the soul, which meant that what you let in would affect it profoundly. You had to learn to shut out the harmful things and let in the good. Though with the world as it is, the priest would probably have added, you had better heed St Anselm. 'Delight coming from the sense is rarely good,' he had once said, 'more often it is truly bad.' [4]

In a sense, the confessional was like a medical dispensary. When Pope Innocent III insisted in 1215 that everyone was now required to confess to their parish priest at least once a year, he wanted Christendom to be purged of what he saw as the sickness of moral laxity. This wasn't just a figure of speech. Care of the soul was important – more important really than care of the flesh. It was only natural then that treatment began with the priest, not with the physician.[5] And if, as some suggest, the Latin term *noxia*, which meant 'harmful things', was at the root of the word 'noise', the implications are clear. Sounds mattered. Some were a potential threat; others were a sign that sickness had already set in. Cackling, angry voices, mocking and bellowing, high-pitched whistling: all these noises were associated with the devil. Blasphemers were often described as 'hissing' their offensive words, rather than just speaking them. The uncontrolled sounds of the mad marked them out as being possessed by demons.[6] Hell, it went without saying, was a noisy place – not least because of the incessant wailing of its inhabitants. Given all this, it was hardly surprising if the priest started his medical mission by getting parishioners to guard their ears and hold their tongues.

If the confessional was a good place for stressing the need to shut out harmful sounds, the main part of the church was a place where parishioners were most definitely required to prick up their ears. The words a priest issued from his pulpit offered a means of salvation for diseased souls. After all, Mary had conceived Jesus at the precise moment she heard God's speech addressed to her by the Archangel Gabriel. The priest offered a workaday version, since he, too, was acting as a channel of divine communication. When people listened attentively to his sermons, they were hearing the word of God. Naturally, what he talked about mattered a great deal. He would have important things to say about scripture, behaviour and morality that needed to be heard. But just as the ringing of church bells had a sacred power, the very sound of the priest's words also had force. It was as if an entire congregation were receiving a blessing and might be cured of their ills if they let his words seep in.[7] That is probably why in the 1280s a woman called Edith, suffering terribly from some kind of madness, was brought in chains to Hereford and placed near the cathedral pulpit: her family would have hoped she was in the best place to be infused with holiness. Medieval religious writing often refers to God's voice as sweet, or perhaps like a wind bringing the Holy Spirit, and touching – indeed, *changing* – whoever heard it.[8]

The sheer presence of godly words, then, would create a special kind of aura. This meant it was especially important to recite them at life-and-death moments. When a Benedictine monk was close to death, for instance, he would have the Gospels read to him day and night. As he slipped away, more monks would come to his bedside and chant the creed, and when his body was laid to rest in the church they would sing psalms continuously in its presence. These, obviously, were moments of particularly intense devotion, but they were only an exaggerated version of the normal state of affairs. Day in, day out, monasteries weren't just echoing with the sound of bells, they were also reverberating with the sound of human voices: reading was usually done aloud, even if quietly. The idea, again, was that this would launch God's words into the air, where they could spread their moral powers to all who were in earshot.[9]

Among God's most zealous followers, medieval ideas about noise could sometimes be taken to extraordinary lengths. Still today, in the Italian town of Guardia Sanframondi, Christian penitents walk through the streets once every seven years, whipping themselves and ritually piercing their skin with pins until the blood flows. It's unsettling enough to watch even now, but in the fourteenth century, when the Black Death ravaged Europe, flagellants would travel from town to town, whipping themselves even more ferociously, wailing, singing hymns and chanting loudly.[10] Although the craze never took off in England, one of the most revealing descriptions comes from the the medieval chronicler Robert of Avesbury, when he described a group of more than a hundred men from Flanders whipping their naked and bleeding bodies during a brief and unsuccessful tour of London in 1349:

> Four of them would chant in their native tongue and another four would chant in response like a litany. Thrice they would all cast themselves on the ground in this sort of procession, stretching out their hands like the arms of a cross. The singing would go on and the one who was in the rear of those thus prostrate acting first, each of them in turn would step over the others and give one stroke with his scourge to the man lying under him.[11]

These flagellants offered the London crowds quite some spectacle. The unsettling sound they were making, though, was just as important as their physical appearance. In seeking penance, it seems they believed that if only enough effort was made, evil might be driven away by the sheer force of humanly made noise.[12]

This is a pretty extreme example of human behaviour, of course. Most of us today would think of it only as a distant echo of a strange and irrational past. Yet even at the time most people regarded the flagellants as absurd. And, as the historian Robert Bartlett points out, we need to be careful not to paint the whole of this period as 'a cartoon "other" to modern pragmatic rationalist society'.[13] In fact, a lot of the medieval attitudes to sound came from a genuine attempt to think carefully about its nature on the basis of the best science of the time. A great deal of rigorous thought was applied, for instance, to harmony, which had been understood in mathematical terms since classical times. The deepest thinker on such matters in the medieval period was probably Boethius, who pulled together the writings of Greek, Arab and Latin philosophers; and especially the work of Plato, to explain the existence of three kinds of music. First there was the 'music of the spheres', made by the cosmos itself as the planets moved through the heavens. It was real and loud and never-ending, but it existed beyond the normal range of human hearing. Then there was 'instrumental' music, the kind that came from, say, trumpets or bells or indeed from voices singing. Finally there was 'human music', the sound generated within each human organism. This bound together body and soul, but like the music of the spheres it was inaudible to our own ears.[14]

Since all these sounds had a mathematical basis in terms of frequency and pitch and so on, it made perfect sense to think that an orderly, godly universe would involve them all being in alignment. The music binding together body and soul, for instance, needed to be in tune with cosmic harmony. Indeed, a person could be thought of as being like a musical instrument that occasionally needed retuning. And instrumental music, though it could lead you astray, might also be thought of as a means of putting things right. Music could, in other words, offer a form of therapy.

Although you might think that the great abbeys and monasteries of medieval Europe are the last places you would expect to find evidence of music being listened to for pleasure, at St Augustine's Abbey in Canterbury, we get a small but significant glimpse of this strict stand against music being broken – at least when it came to treating the gravely ill. At some point in the thirteenth century, new guidelines were issued to the monks running the abbey's hospital. And the wording is quite unambiguous:

> ... for reasons of greater need, if it be judged very useful for improving someone's condition – as when it happens that any brother be so weak and ill that he greatly needs the sound and harmony of a musical instrument to raise his spirits – that person may be led into the chapel ... or carried there in some manner, so that, the door being closed, a stringed instrument may be sweetly played before him by any brother, or by any reliable and discreet servant, without blame ...[15]

Note that the door of the chapel had to be closed. It wouldn't do for any healthy monks to be exposed to the sound of the stringed instrument, no matter how sweetly played. But it's interesting, even so, to find that it was a musician, rather than a physician, who was seen as providing the best form of relief. Elsewhere in the medieval records, we find that having musicians play at the bedside was seen as the most effective response to a quite staggering list of conditions: mania, melancholia, fever, pain, insomnia, plague, lethargy, apoplexy, catalepsy, consumption and epilepsy.[16]

Again, we should avoid looking back at all this and judging it as yet another example of that irrational medieval mindset. We're not so different ourselves, after all. We might buy CDs of Gregorian chants and other gentle music to listen to, perhaps in the bath, in the belief, as the advertising blurb reassures us, that listening to it will restore our inner peace and harmony. There's even a hospital in Missouri that prescribes harp music to its most seriously ill patients. In New York, the neurologist Oliver Sacks confidently asserts that for his own patients, suffering from all sorts of brain disorders and sensory disruptions, 'music has been the profoundest non-chemical medication'.[17]

As for the supporters of musical therapy in the Middle Ages, they knew, just as we usually do, that there are limits to what is possible, even with the sweetest of melodies. In the thirteenth century the whole idea of the music of the spheres was already being questioned; ideas about harmonic realignment were changing. And music wasn't always offered in the hope of some profound metaphysical transformation. Like whispering comforting words into a patient's ear, it was often given in a compassionate attempt to lift the spirits – no more. It's true that in the medieval mind the sound of music – like the sound of a voice – was always believed to be a genuine force, able to change what it touched in some way or another. It could create an environment of goodness or evil; it could heal or infect. Yet behind all the religious dogma and superstition, people knew about allegory or metaphor and were often pragmatic. And all the time, they were learning more and more about the properties of sound and how to fine-tune their hearing.

13

Heavenly Sounds

Walk into any of the great medieval religious buildings that have survived until this day, and you will suddenly enter what is still one of the most distinctive soundscapes you can ever experience. Whether it's one of Christian Europe's cathedrals or abbeys, a Hindu temple in India or one of Islam's grander mosques in the Middle East or North Africa or central Asia, as you cross the threshold and leave the sounds of the city behind, you emerge into an enclosed, cavernous universe of stone. A peacefulness of sorts descends.

Silence, though, is elusive, for large spaces bounded by stone will always reverberate loudly. Cathedrals, temples and mosques can therefore be quite extraordinarily noisy places when people inside are talking, even if quietly, and walking around, even if they've removed their shoes. A door shuts, a prayer book is dropped, a child coughs: these sounds will ricochet around the sheer walls, domes and high ceilings, getting louder as they go. Take two of the largest places of early Christian worship: the Pantheon in Rome and the church of St Sophia in Constantinople. They are certainly visually stunning, but acoustically they are a much trickier prospect. Stone buildings of only half their size would have presented enormous difficulties for anyone wanting to sing or preach and be heard clearly, either by their gods or by their congregation.

Yet somehow in the Middle Ages and the sixteenth and seventeenth centuries performers didn't just make do in these spaces, they found ways of creating sounds which fitted them beautifully. Sometimes there were ways of making the spaces fit the performer, and so the very appearance of the buildings themselves were changed, simply in order to improve the sounds inside. In mosques, of course, the problem wasn't singing or music, which was generally absent anyway, but how to make the words of the Muslim preachers heard. It's why, inside the sandstone courtyard of the vast Babri Mosque which once stood at Ayodhya in India, several large recesses were built into the walls precisely where the speaker would have stood: they would have helped to project his words outwards to everyone at prayer, while also preventing any unwanted echoes. In Christian churches and Hindu temples, by contrast, music or singing was always a vital part of most services. The basic task, however, was much the same: to find the perfect match of setting and voice. In this, both musicians and architects succeeded magnificently. For in the many variations of plainchant and polyphony that emerged in the Romanesque and Gothic cathedrals of the Christian West, and in the devotional music of the Hindu temples of the East, we find some of the most sublime and celebrated music of the entire medieval and Renaissance period.

In this chapter, though, I don't want to offer a potted history of nearly 1,000 years of religious music. Instead, I want to explore something simpler: what it would have been like to be surrounded by these sounds when they were first created. I also want to explore how they encouraged a sense of awe among those who witnessed them at the time. It may be, for instance, that the sound quality of a medieval church makes little sense unless we think of it as a place where ancient and now unfamiliar musical techniques were practised. Sometimes, too, a religious building was used by its rich and powerful patron as a convenient setting for showing off, which means that through careful reconstruction of the past we get a glimpse of how architecture and music were combined, less to create heavenly sound in the service of a god, and more to enhance the reputation and standing of a secular prince.

The first thing to appreciate about medieval Christian churches in the West, though, is how different they would have sounded, even without music, for anyone leaving behind the earthy, absorbent acoustics of their own small, low-roofed home, made of wattle and daub and timber, and entering the spiritual realm of stone and space.[1] It's true that in Northern Europe the earliest medieval churches were themselves made of wood, and here the contrast would have been less stark. But by the twelfth century, most were built of stone. And so back then, if you were to walk into, say, the beautiful Romanesque chapel of the abbey of Vézelay in Burgundy – as thousands upon thousands of pilgrims would have done while breaking their long journey to Santiago de Compostela in northern Spain – the effect would have been stunning. As you entered the nave and saw the rays of sunlight from the windows above fall in shafts on to the pale stone walls and floor, you would have heard the bustling sounds of the street outside melt away. And it would have been hard not to appreciate a feeling of being wrapped within a special aural dimension – a feeling, perhaps, that the sound inside Vézelay somehow matched its austere architectural beauty. The moment you probably fully appreciated this special acoustic quality for the first time, however, would have been when the nuns of Vézelay started to sing during one or another of their daily devotions. At that point, you would have noticed that through a careful control of pitch and volume and pace, their voices didn't get lost – in the way many would have done elsewhere – as they bounced from wall to wall and ceiling to floor in an uncontrolled series of echoes. Instead, what happened could only be described as the singers' voices becoming tuned to the abbey. Rather than reverberant echoing, there was something more like a resonant harmonic sound, in which the sound of the nuns and the sound of the abbey merged into one. The nuns still sing at Vézelay today, so it's possible to hear for ourselves what happens: the abbey amplifies the human voice, but in such a way that instead of scattering it, it focuses it.[2]

How can this be? The answer, it seems, lies partly with the singing, and partly with the building itself. Iégor Reznikoff, the French music historian who, you might recall, analyses the resonant qualities of painted caves, is first and foremost an expert on the music that would have been heard in Romanesque churches such as Vézelay. Indeed, being a trained vocalist, he regularly performs in the abbey as part of his research. On this basis, Reznikoff reckons that the kind of singing originally done in such places would most likely have been conducted, not in what he calls the 'modern equally-tempered scale', but in something he describes as 'just intonation' – that is, in the 'natural pure intervals of resonance'. This, he suggests, would have represented an ancient form of singing or 'modal chant' inherited from antiquity, and would have had its own distinctive musical scales.[3] The essential quality of this ancient chanting in a place such as Vézelay was that it works with the natural echo of the building, rather than against it.

The high notes are gently amplified, and they blend with the lower ones. As the layers of chant are built up on top of each other, they 'lock together' harmoniously, rather than clashing. The effect of using natural intervals is such that even with a single voice, the richness of the harmonics often gives 'the impression of a singing choir, a choir of angels'.[4]

None of this would work if it weren't also for the building's design. As a Romanesque church, Vézelay took its original inspiration from the curved domes of the later Roman world – as did Byzantine church designs, and, through them, the mosques of the Muslim world. Curves were a crucial element of the acoustic design of all these churches: they focused the sound of a singing voice and prolonged its harmonics. Sometimes, of course, one smooth curving dome could be just as much of a problem as a straight wall, simply by being too large, too smooth – and altogether just too resonant. That, essentially, was the Pantheon's problem. But in a slightly smaller, more intimate space, such as at Vézelay, walls and ceilings were treated not as flat or smooth entities but as brilliantly exploitable three-dimensional objects. So the stone vault, which the Romans had used only on the small scale, in places such as crypts, was now exploited for the broadest spans of a church. At Vézelay there was to be a vaulted apse as well as a vaulted nave, and a whole series of rounded arches marking separate bays, all of which created what one architectural historian has called 'a lively and rhythmic space'. Crucially, the presence of all these extra curves and recesses meant any singing sound would have been magnificently concentrated.[5]

It's tempting to translate the fusion of music and architecture we find at Vézelay into a grand edifice of arcane theories about the 'harmonic proportions' of architecture in general. But the notion of abbeys and cathedrals being built according to some fixed and quasi-mystical proportional ratio is misleading. For a start, many church buildings, Vézelay included, were built, adapted and rebuilt many times over, making it difficult to think of them as having a coherent design to begin with. As for the classical works on architecture and geometry, such as the writings of Euclid, Vegetius or Vitruvius, medieval scholars certainly knew about them, and it's significant that Vitruvius, for one, certainly regarded music as part of the essential training of an architect. Yet few medieval master builders would have bothered to learn about this classical heritage. Usually they improvised around taken-for-granted templates: ratios were used, but simply, as Roger Stalley points out, as a 'convenient way of making decisions' on the basis of previous successes rather than in an attempt to endow buildings with divine or mystical connotations.[6] What Vézelay does show us, however, is just as exciting in its own way as any notion of cunning harmonic design: a more organic reverence for sound and atmosphere. Reznikoff refers to it as the 'human scale' of the Romanesque. The soaring Gothic cathedrals that became widespread in Europe from the thirteenth century were undoubtedly more spectacular visually, but, he believes, they were increasingly out of kilter sonically. That delicate balance between visual appeal and acoustic intimacy, which places such as Vézelay had managed to embody, was in danger of being left behind.[7]

Are we in danger of romanticising the Romanesque, and seeing it as a uniquely special period in the history of sound? Perhaps. Reznikoff's work on the ancient modes of singing performed in places such as Vézelay is certainly important, but it mustn't deafen us to the achievements of later medieval and Renaissance musicians and architects, for, as it turned out, they were just as capable of moulding sound and stone together in inventive and spectacular ways. In Europe, just as the new Gothic style of cathedral, with its dramatically higher – and dramatically more reverberant – nave was spreading, we also find the rise of polyphony, a more embellished form of singing, in which instead of a unified sound, different parts of a choir sing with more than one note at a time. It might have started as a casual experiment or even by accident. But in time the best polyphonic singing was adapted to the exact space in which it was performed, so that, with a little practice, the reverberation of a particular cathedral or abbey created just the right mix of overlapping tones and prolonged notes, and instead of acoustic chaos, a kind of soaring celestial harmony was created which somehow matched the building's soaring architecture. The plainchant of the earlier period didn't go away, but it was increasingly marginalised by a profusion of ever more complex and layered styles.[8]

In India, at roughly the same time, musically minded architects managed quite literally to fuse together sound and stone. In some Hindu temples built in the later medieval period, pillars of solid granite were intricately designed to be played as musical instruments. Among those that can still be found, and which tour guides will play for a fee, are several pillars in the Purandhara Mantapa hall built at Hampi in Karnataka during the sixteenth century. These, and other pillars like them, were far more advanced than the musical stalagmites we discovered earlier in the prehistoric caves of Western Europe. In these Indian temples, pillars were individually 'tuned' by means of their length, width and even tension – adjusted by varying the load from above. That way, when played to accompany devotional readings and dance performances, a whole range of different sounds could be deployed, each matched to a different purpose or meaning.[9]

Across India, temples were places which absorbed a host of musical influences – from ancient Vedic chants, mystical sects, even Shi'a Sufi rituals – and then nurtured them as if they were indigenous and distinct. But temples were a lot more than incubators of Hinduism: they attracted vast numbers of pilgrims, they regularly held festivals, and they acted as administrative centres and places of commercial activity. Most importantly, they attracted powerful political patrons, not least because they also symbolised a strong sense of regional identity.[10] In Western Europe the Catholic Church tried to impose complete uniformity in matters of ritual, but here, too, cathedrals and abbeys were objects of local pride. And medieval Europe was a fractious place, politically and culturally – a patchwork of independent lordships, duchies, principalities and kingdoms, each keen to draw attention to what made their territories distinct and special. Again, musical culture and architecture could be deployed as a weapon, so that at times the singing that took place in religious buildings worked not so much in the service of God as in the service of burnishing a local brand.

A striking example of this took place in Venice during the sixteenth century. The city, then an independent republic under the rule of the doge, had already established its reputation for both fine devotional music and brilliant architecture, but it was now the doge's own chapel, St Mark's, which became the main nucleus of creativity. There, under the doge's patronage, architects worked in close collaboration with choral composers to conjure up stunning sound effects: not just beautiful resonance, but startling forms of stereo.

These innovators had to work in a church already several centuries old and with a rich tradition behind it. If you enter St Mark's today, you will notice that, with its 'cross plan' layout, its five interconnected domes on the ceiling, and its glittering veneer of mosaic and marble, it looks a lot more like the Greek churches found to the east, in Byzantium, than most of those in Western Europe. It's stunning to look at. Acoustically, though, it's a complex space. Domes are always challenging, yet having five of them is better than having just one, and the irregular surface of the mosaics prevents quite a lot of unwanted reverberation. Overall, then, it's a good place for music – if not quite an exceptional one.

But when you reach the chancel and the pulpits, you will find some strange sixteenth-century modifications. Two organs are positioned on opposite sides, and inside the chancel, which is screened off from the rest of the church by a rood screen, there are not one but two singing galleries – again, like the organs, facing each other. There can't be anything accidental about this arrangement: the modifications were made under the direction of the doge's architect, Jacopo Sansovino, and his newly appointed musical supervisor, Adrian Willaert – two men already famous in their respective fields. The most likely reason for them has been provided by choral experiments conducted here a few years back by the choir of St John's College, Cambridge.[11] They took one of Willaert's psalm settings and performed it in a number of different ways: sometimes with chant singers behind the high altar, sometimes with a four-part polyphonic choir. When the chant choir was in one of the singing galleries, and a four-part choir in the other one facing them, it created a clear dialogue effect. But when there was a four-part choir in each gallery the sound was even more striking: there was just the right amount of space between the two groups, and this resulted in a beautifully clear stereophonic effect.

Willaert hadn't invented this double choir with its *coro spezzato* (literally 'separated choirs') formation: it seems to have evolved first in northern Italy earlier in the sixteenth century.[12] But what is significant is that in St Mark's it was performed, thanks to the singing galleries, in the very part of the church in which the doge and all his fellow dignitaries now sat as a result of Sansovino's rebuilding: the chancel. In what was now in effect a church-within-a-church, it was the great and good who experienced most fully the drama of this extraordinary musical display.[13]

It's an intriguing bit of musical history. And quite an important one, too, since Venice was a major centre of printing,

which meant that new choral music for split choirs was dis-
seminated rapidly across Europe – in turn triggering a tidal
wave of experiment that culminated in the sacred music of
Monteverdi. But, for me, what is more interesting is the play
of power politics in this acoustic event. Venice was a place of
lavish state ceremonies, all designed to emphasise the mag-
nificence of the republic, its fiercely guarded independence
from Rome, and, of course, the status of the doge himself.
Chroniclers who described the visit of Henri III of France to
Venice in 1574 referred repeatedly to the fanfares of trumpets, the fifes and drums, bells, artillery salutes and singing
that accompanied his tour. It's clear that in these ceremonial
events, sound was being used to express power and status
just as much as the visual spectacle on offer. Everything that
happened in St Mark's – including the experience of listening
there – was carefully constructed to reflect well on the doge
and allow him to bask in its glory in front of any important
visitors.[14]

This pattern of religious music being used first to flatter
and then to represent secular rulers wasn't confined to Venice, of course. Across Europe, princes were competing fiercely
with one another to create a ritual life at court that would
add lustre to their rule. In Constantinople, from the sixteenth
century the sultans of the Ottoman Empire impressed both
subjects and foreigners through the sheer dazzling pomp and
pageantry of their new capital city and their new palace of
Topkapi. Retinues of drummers and horn players, musicians
aboard floating rafts, and explosions of cannon and fireworks
accompanied nearly every royal event: as one historian puts
it, everything here was about 'impressing and overawing with
their displays of wealth and power'.[15] There was plenty of this
pomp in India, too, especially under its own medieval sultans.
Indeed, many of the country's largest temples were funded
and built by rulers keen to project an image of themselves

as semi-deified. The rituals held in these temples were never entirely spontaneous expressions of spiritual or musical creativity, therefore; they would always be required to be dignified enough to reflect well on their patrons.[16]

Religious leaders were obviously wary. In the West, especially, they listened to the increasingly elaborate polyphonic compositions all around, and feared that in many places of worship musical rhythms and harmonies were taking on a life of their own, separate to any devotional content. They listened to the pomp and ceremony of the aristocratic courts and worried that ecclesiastical authorities no longer had a monopoly in the use of sound as spectacle. But really no one had a monopoly, for this wasn't just a struggle between religious and secular leaders. Ordinary people – the peasants of the medieval world – could also make a great deal of noise, and through their dancing and revelry they could, very occasionally, upend the whole social order.

14

Carnival

Twenty-first-century austerity economics, it seems, has created its very own signature noise: the clattering din of metal pots and pans being banged as people march down city streets, gather in squares or even occupy whole neighbourhoods. It was the sound of the millions of *indignados* who filled the centre of Madrid and other Spanish cities in May 2011 to express their frustration at becoming the latest victims of a global economic meltdown. It was heard again a year later when the students of Montreal protested against huge increases in university fees and – after city authorities tried to ban all spontaneous gatherings – thousands of other outraged citizens joined them.[1] And it was also the trademark sound of a coordinated worldwide protest in October 2012 against the role of banks in creating debt, when 'casserole' protests erupted in the streets of Lisbon, London, Madrid, New York, Istanbul and over a hundred other cities stretching from Argentina to Japan.[2]

During these protests, a turbulent cacophony of issues and slogans were heard, each demonstration clattering to a slightly different tune. But using pots and pans in every location – creating a shared global sound – is both deliberate and symbolic. Almost anyone can join in, since the musical instrument of choice is simple and cheap. It's meant to show that everyone's grievance is somehow connected. Perhaps most potently of all, it's a noise that draws on a history of disruptive music-making by the oppressed that goes back more than 700 years. For today's 'casserole' marches are just the latest version of a rich tradition of carnival and protest that had its origin in the Middle Ages.

Those two words – carnival and protest – really do belong together. Making noise in this way, with people gathered together in rhythm and spirit, can obviously be fun and invigorating, yet history tells us that noise also stirs emotions, it intimidates. Sometimes it's meant to intimidate; at other times, the threat is purely in the ear of the beholder. In the Middle Ages, and in the sixteenth and seventeenth centuries, there wasn't always a clear line between revelry and revolution: the unsettling reality was that very often they sounded exactly the same.

A millennium ago, on local saints' days or other special dates in the religious calendar – Epiphany, Ascension, Pentecost, Corpus Christi – thousands of parish churches all across Western Europe could easily have been filled with the sounds of dancing, music, singing and feasting. Christianity, we should recall, started out as one of Rome's eastern 'mystery' religions: among its earliest followers were plenty who believed that losing themselves in ecstasy through dancing and singing was the best means of experiencing the divine.

For centuries afterwards, church services might have regularly featured parishioners and priests singing and dancing together – as in the celebrations for St Eluned's Day, taking place at one church in Wales and described by a twelfth-century traveller:

> You can see young men and maidens, some in the church itself, some in the churchyard ... in the dance which wends its way round the graves. They sing traditional songs, all of a sudden they collapse on the ground, and then those who, until now, have followed their leader peacefully as if in a trance, leap up in the air as if seized by frenzy.[3]

Naturally, official Church thinking disapproved of this sort of thing, which meant a long – and not always entirely success-ful – campaign to make churches more respectable, orderly places. If parishioners' senses needed stimulating, the argu-ment went, let it be done through more disciplined singing, more elaborate rituals, through incantations, incense and bells. All the wilder stuff – any dancing or ecstasy that people might crave – must be restricted to special occasions, kept in the churchyard, maybe pushed on to the streets outside.[4]

Of course, once outside, playfulness could run away with itself. Festivities were still under the watchful eye of the Church, yet no longer under its direct control. People were free to start creating their own sense of what it meant to step out of their everyday lives and have a little fun. True, there was still a lot of religion and civic pride about, but there would be more profane pleasures, too.

So on a feast day or holy day any time between the four-teenth and seventeenth centuries, we would have heard mixed together in the streets of almost any European town the sounds of mystery plays re-enacting biblical stories, lively processions through the town featuring local civic dignitar-ies and representatives from various craft guilds – tanners, coopers, hosiers, butchers, bakers, and so on. There would

probably have been some noisy rivalry between the different
guilds as each one competed to put on the most striking display. There would certainly have been games and sports such
as archery or wrestling or animal baiting. And a deaf ear would
have been turned to behaviour which, on any other day of the
year, would almost certainly have got ordinary people into
deep water: masters swapping roles with servants; men dressing up as women and vice versa; ribald songs poking fun at
local bigwigs; perhaps someone got up as a Lord of Misrule,
whose duty it was to ensure a steady stream of buffoonery and
subversion.

Even parts of the religious establishment acknowledged a
real social need for all this. 'Foolishness', the School of The-
ology in Paris declared in 1444, was second nature in all of us.
It must 'freely spend itself at least once a year', for after all,
'Wine barrels burst if from time to time we do not open them
and let in some air'.[5] This, in other words, was carnival as an
officially sanctioned – though strictly time-limited – opportunity for everyone to let off a little steam. As E. P. Thompson
explained, 'Many weeks of heavy labour and scanty diet were
compensated for by the expectation (or reminiscence) of these
occasions, when food and drink were abundant, courtship and
every kind of social intercourse flourished, and the hardship of
life was forgotten'.[6]

Apart from being noisy, a carnival was also obviously a
pretty boozy affair. Huge quantities of beer and wine were
usually brought in for the occasion and then either sold dirt
cheap or allowed to flow freely through conduits in public
squares. Perhaps it was inevitable that things sometimes got
out of hand, town centres sliding into drunken, debauched
and violent no-go areas. It was in volatile circumstances like
this that the opportunity might be taken to settle old scores –
everything from poking fun at notorious individuals who had
broken social conventions to full-blooded insurrection.

Matters certainly ran out of control in the small Somerset cathedral city of Wells in May 1607. That year, amid the usual round of springtime games and feasting and festivities, one group of parishioners decided to organise a 'church ale' in the city to raise some money for repairs to their church, St Cuthbert's. It was the kind of charity event that usually lasted a day or two, but in 1607 the good citizens of Wells managed to stretch it out rather impressively: from May Day through Ascension Day, Whit Sunday, Trinity Sunday and St John's Day, all the way to the 25th of June – a grand total of eight weeks.[7]

Instead of a sedate picnic of cakes and cream and roasted meat washed down with a little freshly brewed ale, organisers laid on the full panoply of popular seventeenth-century entertainment: plays and burlesques from every neighbourhood and trade guild; dressing up as a May Lord and his court; re-enacting the tales of St George and the Dragon and Robin Hood and his outlaws; morris dancing – the list went on, as, of course, did the drinking and eating.[8]

Throughout the eight weeks, the noise drifting over the rooftops of Wells barely stopped. There was street dancing, with all and sundry following the May Lord and his Lady and accompanied by fiddlers, horn players and drummers – weaving their way, as one witness put it, 'with muche hooping and hallowing'.[9] Trumpets blasted away at regular intervals – sometimes, rather mischievously, simply in order to disturb the night watch or taunt the bishop in his palace. Another sound, which kept up almost non-stop through the eight weeks, was the beating of drums. There were drumbeats to call the revellers together, drumbeats for the dancing, and drumbeats for the parades. They carried on late into the evening, and sometimes began as early as four in the morning. No one, it seemed, had hired the drummers; they played, they said, purely for personal pleasure.

At regular intervals that summer, Wells also resounded to the raucous din of a 'skimmington ride', or a charivari. It was then that the so-called 'rough music' would begin. This was a tradition that stretched right across Europe, with local variations in France, Italy, Spain and Germany. The general idea was that groups of people, usually young men in disguise, would gather at night, often armed with pots and pans and cowbells, and then set off to harass some poor person – a widow or widower, perhaps, who had remarried someone scandalously younger. The rabble-rousers would stand outside his or her house and, with a loud clattering, and a great deal of singing and shouting and roaring of animal noises, they would stir them from their slumbers and taunt them, perhaps for hours.[10]

The reputation of the medieval charivari is of a boisterous but essentially genial affair, in which those being harassed would be left alone after paying some kind of fine: an inclusive occasion for bonding, and a way of enforcing community standards without resorting to violence.[11] It had, however, a darker side. It's not hard to imagine that when a community set out to enforce its standards in this way, anyone different or unusual who had become a subject of gossip and prejudice could quickly become a target. Frankly, these escapades were often an occasion for bullying.

By the sixteenth and seventeenth century, they were also taking on a more political dimension. In Wells, revellers targeted the city's civic and business leaders – and, above all, the Puritans in their midst, who had tried to suppress the festivities right at the start. Some of that festive drumming was designed very deliberately to drown out attempts to disperse the revellers. The morris dancing wasn't all sweetness and joy, either. Performers would often ride hobby-horses and jerk them about in a series of aggressive thrusts, or carry sticks and shout and move their bodies in a choreographed version of combat.[12] Then there was the gunfire. One of the men targeted by the revellers later accused them of being 'armed with unlawful weapons'.[13] It's not clear whether this witness was embroidering the truth, but it certainly wasn't unknown for guns to be heard at such gatherings, and they had been putting in an appearance at carnivals in France since the sixteenth century. By the 1600s all across Europe they were carried with pride by men processing through the streets and occasionally fired in multi-gun salutes. Just eleven years after the revelry at Wells, and just a few miles away in Wiltshire, there was another skimmington, in which one contemporary account claims 300 or 400 men armed 'with pieces and other weapons' unleashed a volley of gunshots, while others gathered there sounded their pipes and horns or rang their bells.[14]

To those on the receiving end, cowering in their manor houses and rectories, all this must have been pretty unnerving. The chaotic nature of the noise, as well as its sheer volume, represented a calculated attack on the harmony expected of social relations. It wasn't surprising, then, that instead of being tolerated as a means of enforcing social norms, all the disguises and the masks, the banging and the shouting, began to feel more and more to those at the top of society like a cover for rebellion by those at the bottom. Sometimes they were right too, for the uncontrolled noise of carnival certainly spoke, not just of pleasure and abandonment, but also of underlying discontent – anxiety over plague, a weariness and anger over excessive work demands, price rises for bread or just the grinding poverty. It's probably why one carnival in Udine in Italy ended with twenty palaces ransacked and up to fifty aristocrats and their retainers killed. Even if there was no outright rebellion, ordinary people in this period of history were often able to express their political frustrations only in the coded words and songs of festival – through a language of oaths, toasts, seditious riddles, ballads, even 'airs whistled in the streets'.[15]

Increasingly, the noise of revelry was simply assumed to be the noise of outright rebellion. Indeed, sometimes leaping to such a conclusion provided the best excuse for violent suppression. In the southern French city of Romans, for instance, the Mardi Gras festivities of 1580 ended in ambush and disaster. In the thick of the celebrations, as the citizens danced aggressively with swords, brooms, and the flails they usually used for threshing wheat, the city elite lost their nerve and snapped. One man, a draper and the leader of the city's most popular faction, was assassinated. The leader of the ruling party's most reactionary wing then hired a mob of thugs to pursue and beat up the murdered man's friends and supporters. Whatever spirit of revolt the people of Romans had been harbouring was savagely crushed. It was a sign that all across Europe those in authority just couldn't shake off the feeling that there was something deeply threatening in the hubbub and petty disobedience of carnival. Nor was this attitude confined to Europe for long. As countries such as Spain, Portugal, France and Britain colonised the Americas, Africa and beyond, both the carnival tradition itself and the anxiety it caused among elites was exported.

A modern-day example of this is the *pancadão* (meaning 'big punch' in Portuguese), one of the many rhythms that today make up the rich 'sound salad' of the Brazilian favela, or shanty town. The *pancadão* are street parties which take place when three or four cars are parked close together, blasting music from their speakers until the early hours – or at least until the police shut them down. They attract thousands of youngsters, who turn up to 'flirt, drink, and dance to the sound of Brazilian funk'.[16] It's not quite Latin American carnival as we imagine it, but like those clattering casserole demonstrations, the Brazilian *pancadão* are just as much in the tradition of the medieval European carnivals as the much bigger, glossier displays put on in places like Rio de Janeiro.

There's no universal model of what carnival should be, of course. And one of the things that make Brazilian versions distinctive is that the religious festivities brought to the country by the Portuguese colonisers of the sixteenth and seventeenth centuries were fused together with other influences from Africa, as well as mythic indigenous traditions. In the nineteenth century, in Recife and its neighbouring city of Olinda, for example, former slaves of African heritage who had been working in the sugar plantations brought to the Portuguese carnival tradition 'frevo' music, which turned European marching tunes into syncopated rhythms and improvised fanfares. Processions also started to include lancers, who represented warriors possessed by indigenous or African spirits and who danced, leaped and duelled with each other in mock combat, ringing the large cowbells on their backs as they moved about. As in old Europe, there are still lots of carnival clubs, each representing a different part of the community and each competing with its rivals over who has the most dazzling costumes or floats. Yet the clubs are based not just on colonial-era trade guilds or religious factions; they're also based on the so-called 'nations' into which slaves had once been organised by plantation owners according to their real – or assumed – tribal origins.

What the organised carnival processions of Recife and Olinda do share with the *pancadão* of the favela, the casseroles of Quebec, and all their European ancestors, is the participants' desire to step aside from their normal routines of work and let themselves go for just a moment. It's no coincidence that it's the poorest who are the most enthusiastic about these events, which is probably why today's gatherings also frequently share the same fate as those in Europe 400 or 500 years ago. In Brazil, carnival is often a cue for the well-off to leave the city for the quiet of the countryside. They can't help but be suspicious: for them, as it was for the Puritans of Wells back in 1608, carnivals and street parties are an expression of vulgar popular culture teetering on the edge of violence.

Most of this, of course, is – and always has been – paranoia. But in one respect, the rich are right to be wary. The drumbeat, fanfares, singing and shouting of carnival has always been the authentic sound of the dispossessed – of those who feel they have never been heard enough and take whatever chance they have of giving their masters an earful. It has been their way of saying, 'I'm here, I exist, I won't be ignored.' That has never been a comfortable message for the great and good, which is, perhaps, one of the reasons why, in the sixteenth and seventeenth centuries, so many of them started proclaiming more loudly than ever the virtues of good manners, emotional restraint and sealed lips.

15

Restraint

In a middle-class, God-fearing English home of the seven-teenth century, you would be hard-pressed to find much fun or relaxation at the family dinner table – least of all if you were a teenage girl being trained into respectability. For you, especially, the evening meal was a forbidding testing-ground of character:

> Gentlewomen ... keep your body strait in the Chair, and
> do not lean your Elbows on the Table. Discover not by any
> ravenous gesture your angry appetite; nor fix your eyes too
> greedily on the meat before you ... do not bawl out aloud for
> any thing you want; as, I would have some of that; I like not
> this; I hate Onions; Give me no Pepper: But whisper softly to
> one, that he or she may without noise supply your wants ...
> Eat not so fast, though very hungry ... Close your lips when

you eat; talk not when you have meat in your mouth; and do
not smack like a Pig, nor make any other noise which shall
prove ungrateful to the company ... Fill not your mouth so
full, that your cheeks shall swell like a pair of Scotch-bag-pipes
... It is very uncomely to drink so large a draught, that your
breath is almost gone, and you are forced to blow strongly to
recover your self: nor let it go down too hastily, lest it force
you to an extream cough, or bring it up again, which would
be a great rudeness to nauseate the whole Table ...[1]

No doubt there's a lot in that advice from 340 years ago for
which most parents today would raise a glass of their own.
But in 1675, when it was published in *The Gentlewoman's
Companion, Or, a Guide to the Female Sex*, the stress placed on
children avoiding *all* unnecessary chatter at mealtimes might
make even the strictest disciplinarian among us quail:

Be not talkative at Table, nay, nor do not speak, unless you
are askt a question.[2]

The trouble was that the same standards of behaviour were
expected all day long. During the hours of education, for
example:

Neglect not what you are to do by vain pratling in the
School: make no noise, that you may neither disturb your
Mistress, or School-fellows ... In the intervals of School-
time, let your recreation by pleasant and civill, not rude and
boisterous.[3]

Or when going with your parents to church:

Do not proclaim publickly to the whole Congregation your
levity and vanity by laughing, talking, pointing with your
finger, and nodding ...[4]

Even last thing at night:

> Going to bed, make no noise that may disturb any of the
> Family, but more especially your Parents.[5]

Indeed, the guiding principle for all children, but for young
ladies in particular, was very clear:

> It is proverbially said, *Maids should be seen, not heard*; not that
> they should not speak, but they should not be too talkative.
> A Traveller sets himself out best by discourse, but a Maid is
> best set out in silence.[6]

Of course, you can bet silence didn't really reign in most
family homes. In the seventeenth century, households were
often busy places, with lots of coming and going. They could
be businesses: a farm, a workshop, mill, store or tavern. They
were often cramped, occupied not just by parents and their
children but also, perhaps, by elderly widows and widowers,
servants, even young people taken in from other families. And
there's plenty of evidence that even the most puritanical parents didn't always rule with a rod of iron, indeed that the relationship between them and their children was usually 'warm
and affectionate'.[7] Governesses, too, were told by *The Gentlewoman's Companion* that they were not to be 'harsh in expression, nor severe in correcting'.[8] The self-control required of
the young was also required of everyone else.

The Companion's message of unceasing modesty and
restraint in all things would have found a wide readership
among the gentry and merchant classes – people wanting to
make their way up in the world. In the sixteenth and seventeenth centuries, all across Europe and in the new colonies in
America, there was a new emphasis on self-discipline in everyday life, and with it a revulsion against noise of every kind.
It wasn't just the boisterous spirit of medieval carnival that
was tamed, as city authorities and religious leaders everywhere
trimmed back the number of saints' days in the social calendar,

banned singing and feasting from public squares, tore down maypoles and ordered unruly street parties to be replaced by more sedate processions of holy relics.[9] The reaction against unwanted sound went much, much wider. A whole new set of social rules was in circulation about how and when to speak, about laughing, eating, dancing, music, what you should listen to and what you shouldn't. In almost every case, the overriding virtue being preached was restraint. In England, Elizabethan writers explained in graphic detail how spitting, snorting and breaking wind were beastly acts which though once indulged freely were now best avoided.[10] By the eighteenth century, English writers found it difficult even to mention such vulgarities, and in 1748 Lord Chesterfield had the following advice:

> Frequent and loud laughter is the characteristic of folly
> and ill-manners; it is the manner in which the mob express
> their silly joy at silly things ... there is nothing so illiberal,
> and so illbred, as audible laughter ... not to mention the
> disagreeable noise it makes.[11]

As the historian Keith Thomas has shown, the style required to establish one's superiority was now 'gravity and stateliness': rehearsed gestures, euphemisms, a deathly decorum. The same ethos had already transformed dance music. In the sixteenth century the kicking legs and flailing arms of country folk-dances were steadily tamed, to re-emerge in fashionable society as the more cultivated and courtly galliard or pavane. People had once moved together arm-in-arm in freewheeling circles or lines; they now faced each other as couples or individuals, standing ramrod erect and linked by only the briefest of touches. As for polite conversation between consenting adults, the strictures of the *Gentlewoman's Companion* were once again crystal clear:

Be not guilty of the unpardonable fault of some, who never think they do better than when they speak most; uttering an Ocean of words, without one drop of reason; talking much, expressing little …

When you enter into a Room by way of Visit, avoid the indiscretion and vanity of a bold entrance … do it quietly and civilly. And when you come near the person who you would salute, make your Complement and render your Devoir modestly, and with some gravity, shunning all bawling noise or obstreperousness …

Volubility of tongue … argues either rudeness of breeding, or boldness of expression. Gentlewomen, it will best become ye … to observe, rather than discourse; especially among elderly Matrons to whom you owe a civil reverence, and therefore ought to tip your tongue with silence …

To persons of Quality in a higher rank than your own, be very attentive to what they say, lest you put them to the trouble of speaking things twice. Interrupt them not whilst they are speaking, but patiently expect till they have done … If you find the company more facetious and witty than your self, leave the discourse to time, and be silent, contenting yourself to be an attentive hearer …[12]

Why did it happen, all this apparent crushing of spontaneous expression and communal fun? Was it a result of dogmatic religion? A form of class-warfare? Or simply a natural reaction to the world itself becoming a louder place in which to live?

There were certainly lots of complaints about rising levels of noise well before the Industrial Revolution. True, poems and songs of the sixteenth and seventeenth centuries still rhapsodised about the rural charms of birds singing in the forest or meadow, the splashing of streams, the trees whistling in the wind; while big-hearted souls wandered among the bustling shops, taverns and street hawkers of a fast-expanding city such as London and, like one John Evelyn, wrote enthusiastically about how 'as mad and lowd a Town is no where to be found in the whole world', or, like Orlando Gibbons, Thomas Weelkes and Richard Dering, turned its rich spectrum of cries and fancies into elegant music for posterity.[13]

The reality of daily life in a crowded town or city was grimly different, however. Imagine, for instance, what it would have been like to live next to a musician practising one of Gibbons' new compositions. In Oxford in 1610, a second-floor room was leased to just such a tenant, one John Bosseley. The city authorities were fleet-footed enough to stipulate that he could only practise between certain hours, but his main trade was a dancing school and years later we find councillors insisting that he must not 'daunce nor suffer any dancing after tenne of the Clocke in the night nor before five of the Clocke in the morning'.

It could have been worse. In a compact market town with narrow streets like Oxford – a place at this time still twice the size of Manchester – workshops and cottage industries were often jumbled up with people's homes, and, as the law courts show, the hammering of blacksmiths, the clatter of cutlery-making and the din of late-night revellers spilling out of alehouses were a source of constant irritation.[14] Petitions complained of 'intolerable and continual' noises or of the 'odious' din of drunkenness. And legislators responded. An Act of 1552 required pub landlords to guarantee they would prevent 'hurts and troubles … abuses and disorders' before they got a licence. Another passed in 1595 stipulated that in London, 'No hammar man, [such] as a Smith, a Pewterer, a Founder, and all Artificers making great sound, shall not worke after the hour of nyne in the night, nor afore the houre of four in the Morninge.'[15] The same curfew applied to domestic noises. No man, the law said, should cause any 'out-cry' by beating his wife or servant – at least not after nine o'clock in the evening.[16]

It's difficult to know if life really was getting noisier or people were just becoming more sensitive. It was probably a combination of both, but either way it was impractical for secular authorities to detect and punish every breach of the law. On a daily basis, good conduct depended on the instinctive observance of social convention, and that, in turn, reflected a deeply rooted religious morality about sound. A domestic guidebook such as *The Gentlewoman's Companion* was infused from beginning to end with the idea that people could be improved, moulded, refashioned, and perhaps the main force acting upon them was what they saw and heard around them. In the *Companion*'s pages that ancient fear of sensual corruption was alive and well:

> Let there be a strict watch to keep unviolated the two gates
> of the Soul, the Ears and the Eyes; let the last be imployed
> on good and proper Subjects, and there will be the less fear
> that the Ears should be supriz'd by the converse of such who
> delight in wanton and obscene discourses, which too often
> do pleasantly and privately insinuate themselves into the
> Ear, carrying with them unwholesome air which infects and
> poysons the purity of the Soul.[17]

Listening *and* talking are both in the dock here. Ears should be shut to bad influences, but careless talk can also be devastating. 'As by good words evil manners are corrected,' the *Companion* said, 'so by evil words, are good ones corrupted.'[18] It's a moral code expressed in this instance in Anglican England, yet we would hear much the same idea expressed in Puritan New England – or, indeed, a little later, in Wahhabi forms of Islam. Religion in general took sound seriously as a potent moral force. In New England, preachers regarded godly speech as something able to chisel away at the sinfulness of the world. The more you heard of it, the less sinful you would be.[19] Other kinds of speech – murmuring, grumbling, whispering – were labelled as 'clamourous' and treated as unruly. It's why some New England Puritans were wary of so-called 'Ranters' or 'Singing Quakers'. For them, singing was a cunning display. A preacher would warn his own congregation to keep well away from earshot.[20]

Once again, it was music and song that most agitated religion's more puritanical believers. In seventeenth-century Arabia, in common with the rest of the Muslim world, music and singing could be heard in some mosques just as much as prayer. The eighteenth-century Muslim theologian Muhammad ibn Abd al-Wahhab, though, introduced the idea of purging Islam of ecstatic forms of worship involving dancing and singing and chanting, which he saw as corrupt. So, just as English Puritans had pulled down maypoles, followers of Wahhab expelled music from the mosques. Other branches of Islam, notably the Sufi tradition, held firm to the idea that music could be a means of experiencing the divine. But across large parts of the Middle East, as in the West, even among more moderate sects there was more and more stress on 'the struggle for holiness within the individual soul' – a struggle, in other words, all about self-restraint, about a turning away from the distractingly pleasurable.[21] Even *The Gentlewoman's Companion* rails against 'lascivious and wanton' ballads and a 'contagion of loose Songs', warning that 'they may please the ear, yet may corrupt good manners':

> To take delight in an idle vain Song without staining your self with the obscenity of it, is a thing in my mind almost impossible; for wickedness enters insensibly by the ear into the Soul, and what care soever we take to guard and defend our selves, yet still it is a difficult task not to be tainted with the pleasing and alluring poison thereof.[22]

Religion sorted listeners into the godly and the wicked. More and more, though, people were also being sorted into a *social* hierarchy, depending on how they listened and how much noise they made. Loud laughter, lascivious dancing, loose talk, lewd songs: all these were, like drink, welcome means by which the poor might reconcile themselves to the harsh realities of their lot. But such noisy activities wasted valuable labouring time and poked fun at the rich. Indeed, they had a rebellious quality which, frankly, needed dealing with. So when Lord Chesterfield condemned laughter as 'ill-bred' and *The Gentlewoman's Companion* declared that 'you can never be truly accomplished till you apply your self to the Rule of Civility',[23] both were conforming to a broader pattern: linking the sounds of popular pastimes and spontaneous self-expression with the vulgar, even the bestial.

But to what ends? Should the better sort simply accept that noisier, less discriminating souls – the poor, women, children, slaves, servants – were just different, best kept at arm's length and out of earshot? Or should they try to stamp out these differences, force them into the realm of polite society on polite society's terms, or at the very least make them bearable as neighbours? And what of the people themselves? Did they wish to keep quiet in order to get on – or would they prefer to embrace and enjoy their own ways of life?

There was no simple answer. But increasingly, it looked as though the cult of decorum had split wide open simmering divisions between elite and folk culture and erected new barriers between the classes. By now, in France, some towns were beginning to organise two separate carnivals: one for the restrained middle classes, another for the unrestrained poor. While the working classes persisted in their hellraising, among the upper classes even wedding feasts were becoming private, solemn affairs: they no longer included the whole neighbourhood, just the immediate family. As Robert Darnton points out rather nostalgically, there was 'no more drunkenness, no more brawling at table, no invasions from a rowdy counter-ceremony ... or bawdiness exploding from a charivari or cabaret'.[24] In Britain, whole new towns were soon being built to give the better-off a chance to escape the din of the congested, noisy city centres – and leaving the poor to their crowded, dilapidated tenements. And as forms of entertainment diverged – operas and balls for some, fairs and revels for others – the novelist Henry Fielding suggested that the English 'So far from looking on each other as brethren in the Christian language ... seem scarce to regard each other of the same species'.[25] In America, meanwhile, white European settlers were leaving their New England colonies, attempting to tame – or silence altogether – the alien, wild and apparently threatening sounds of Native Americans and African slaves.

The sixteenth and seventeenth centuries did not fall silent. But by the eighteenth century it sometimes seemed as if different sound worlds were forming – not so much between different parts of the world, but, more fundamentally, between those with power and those without. As the world stood on the brink of the modern age, it also faced the prospect of countless struggles for supremacy between these two opposing groups, in all their manifestations of gender, class and race. And while some of these would be settled peacefully or would simply dissipate, others would explode into full-scale revolution and violent suppression.

IV
Power and Revolt

16

Colonists

In late July 1609, a ship called the *Sea Venture* had almost completed its maiden voyage across the Atlantic when it ran into a storm so violent and terrible that accounts of it are thought to have later inspired Shakespeare's *The Tempest*. The *Sea Venture* had set off along with eight other ships from Plymouth in England a month before to bring badly needed supplies and people to Jamestown in Virginia, where one of the first English colonies in America was slowly starving to death. Just a few days from its destination, as William Strachey, one of those aboard, recalled, the wind started 'singing and whistling most unusually':

> a dreadfull storme and hideous began to blow from out the
> North-east, which swelling, and roaring as it were by fits,
> some houres with more violence than others, at length did

beate all light from heauen; which like an hell of darkenesse
turned blacke vpon vs ... For four and twenty houres the
storme in a restlesse tumult had blowne so exceedingly, as
we could not apprehend in our imaginations any possibility
of greater violence, yet did wee still finde it, not only more
terrible, but more constant, fury added to fury ... our
clamours dround in the windes, and the windes in thvnder.
Prayers might well be in the heart and lips, but drowned in
the outcries of the Officers: nothing heard that could giue
comfort, nothing seene that might incourage hope.[1]

The ships, unable to communicate with each other by the
usual means of lanterns, flags, shouting or trumpets, were
quickly scattered. The *Sea Venture* itself, along with the 150
men, women and children in the hold, ended up stranded on
one of the uninhabited islands of Bermuda. They were alive,
yet as far as William Strachey was concerned, far from having
found a tropical paradise they were now marooned in a 'dangerous and dreaded' place – a place of alien, ominous noises:
'such tempests, thvnders, and other fearefull objects are seene
and heard, that they be called commonly *The Devils Ilands*'.[2]

Even more threatening than rain and thunder, however,
was the risk of anarchy. Some of the survivors were soon arguing that the authority of the expedition's leaders, Sir Thomas
Gates and Sir George Summers, automatically lapsed with
the ship's wreckage – that they were 'all then freed from the
gouernment of any man'.[3] And the simple truth was that neither of the two officers had any military means of enforcing
their power. Gates, however, had the idea of putting the *Sea
Venture*'s bell to use. Every morning and evening, 'at the ringing of the bell', everyone was summoned to public prayer and
a roll call. Those who didn't respond 'were duly punished'.[4]
As the historian Richard Rath has noted, in 'a land with no
churches or courts', the wrecked ship's bell, calling all within

earshot together, 'was the adhesive Gates needed'. Its sound brought a sense of order and familiarity. It created an invisible bond between the different families. And it established both the source and the geographical reach of all power on the island.[5] When, many months later, the families managed to leave Bermuda and land at Jamestown, Governor Gates's very first act was to march straight to the church and order its bell to be rung.[6]

We have already learned how, for thousands of years, bells had been rung in China, India, the Middle East and across Europe to draw people to prayer and to send evil spirits fleeing. But in 1609 the sound of the bell became part of another story: the story of colonisation. It helped the most powerful nations in the world to overpower alien soundscapes, create order out of what they thought of as chaos, mark their territory and, in time, build empires. The bell was only one modest weapon among many for the colonisers. There were also drums, trumpets, pipes, horns – and, of course, guns. Armies and navies – and the deadly firepower they had – certainly did all the dirty work of conquest, yet the sounds of colonialism mattered too. The settlers' guns were not just deadly, they were extremely loud – frighteningly so for those who had never come across them. Like the bells and drums and trumpets, the noise that guns made helped to establish rule over people whose lands were seized. They also helped instil discipline among the new communities they forged. Indeed, in places such as America and, later, Australia, it was sound that would often prove to be the best way of making the presence of a distant authority felt – and, very often, feared.

The model for what could be achieved had already been worked out in the Old World, where in the sixteenth and seventeenth centuries the administrative machinery of the nation state had been busily trying to extend its political reach outwards, from the royal court at the centre to the remotest corners of the countryside. The Habsburgs of Vienna, the Bourbons in Versailles, the Ottoman sultans of Istanbul, the Mughal emperors in India: all these dynasties wanted not just their bureaucratic writ to run throughout their lands, but also their authority – their divine right to rule – to be *experienced* on a daily basis by all their subjects. Nation-building on this level, encouraging or cajoling people into a sense of belonging emotionally, was often achieved through symbolic displays of royal authority: pomp, pageantry, painted images, festivities, processions and uniforms. But the symbolism wasn't always visual. In every parish, loyalty to the crown could be instilled orally, through sermons from the pulpit.

Royal authority also reached almost every town and village through the sounding of the post horn. Here was a carefully regulated instrument of the state in more senses than one. It deployed a precise code of signals to indicate the arrival of different sorts of mail – express, normal, local, packages – as well as different calls for the arrival or departure of post riders and mail coaches, or to indicate distress. There were even special calls to warn any changing station ahead of the number of carriages and horses required.[7] And, as one later worker for the Austrian postal system described, the instrument had a certain romance attached to it:

> Through the narrow streets and across the country landscape the post horn was heard, in the villages and alleys of the cities, at the gates of castles above and by the monasteries below in the valleys – everywhere its echo was known, everywhere it was greeted joyfully. It touched all the strings of the human heart: hope, fear, longing and homesickness – it awakened all feelings with its magic.[8]

The post horn would also, perhaps, have prompted among all who heard it an intuitive respect for the distant civil authority upon which it depended – and which in turn it audibly advertised. Its sound was, in a real sense, the sound of governance, the idea of a realm of peace and order, and a way of marking out its territory.

In the New World it made sense to import traditions like these to help hold together the precarious colonial settlements taking root in places such as Jamestown in Virginia or, further up the coast, in Massachusetts and Connecticut. When the very first settlers had arrived in Jamestown in 1607, two years before the *Sea Venture* set sail, they had let loose a fanfare of trumpets. It somehow marked – legitimated, even – their capture of the land for both colony and king.[9] Later, they rang bells to muster troops as well as congregations or to summon people to a town meeting. If they couldn't afford bells, they would make do with drums or conch shells. When leading members of their community died, they would launch a volley of gunshot and cannon. As Richard Rath has shown, it's now that we start to find the term 'earshot' being used in print. It defined the limits of a colonial community, just as the range of a church bell had long defined the limits of a parish. However, guns and cannons were even louder than bells: they extended the reach of civil order, and wherever they were heard, they conveyed not just the will of the colony's governor but also the power and divine right of the crown 3,000 or so miles away.[10]

The colonists often liked to imagine their new homeland as an empty wilderness, but of course they weren't the first to live there. Jamestown was on land that had belonged to the Powhatan people. Further up the coast, colonists moved on to the territory of the Iroquois, the Algonquin, the Huron and other Native American groups. One striking feature of the uneasy relationship between these established inhabitants and the European newcomers was how they described each other as sounding so different. Take, for instance, the blood-curdling account of Mary Rowlandson, briefly captured by the Narragansett in 1675. She later described how 'a company of hell-hounds, roaring, singing, ranting and insulting' had dragged her and some of her children away. Their first night in captivity was terrifying:

> Oh the roaring, and singing, and dancing, and yelling of those black creatures in the night, which made the place a lively resemblance of hell ...

Later, Rowlandson overheard another group of Native Americans approaching:

> Oh, the outrageous roaring and hooping that there was! They began their din about a mile before they came to us. By their noise and hooping they signified how many they had destroyed.

These sounds must have been intimidating. Yet, even after one of her children had died, she was keen to use her account to show that she had clung to her own world and its faith:

> I have thought since, of the wonderful goodness of God to me, in preserving me so in the use of my reason and senses. In that distressed time, and that I did not use wicked and violent means to end my own miserable life.[11]

She was godly; her captors, as one of her fellow colonists put it, were 'hellish'.[12] The Native Americans' language, their prayers, their songs: these were all reduced by Mary Rowlandson and her fellow colonists to 'foule noise' or meaningless 'howling'. Though actually it was never completely meaningless: for Christian Europeans, these unfamiliar 'hellish' sounds were a vivid measure of the Native Americans' wildness and savagery.

Naturally, the Native Americans' own perspective was altogether more complex. For most of them, sound itself was alive. When a clap of thunder was heard, an identity was attributed to it. By extension, to be silenced as a human being was tantamount to losing one's self. That is why among the Iroquois it was important when under torture to keep singing your own tribal 'death song' for as long as you could bear. If you could do that – remain in control of your voice, avoid all involuntary wails or whimpers when suffering horribly – then you remained unbroken, even if, at the end of it all, you died. The same thought propelled Native American shamans to keep up their drumming and singing: it was what enticed the birds or fishes or animals towards them; and whenever Christian missionaries silenced the shamans, they claimed their hunting ceased to be successful.[13]

So colonial life in America involved a struggle for supremacy in which sound was both an offensive weapon and something precious to be defended. Colonists thought of the noise they themselves made, like God's, as being 'like Thunder-bolts, knocking down all that stood in their Way'.[14] The Native Americans' wild demonic noise might therefore be counter-acted with a controlled, godly array of sounds, which, as William Strachey explained, meant that deadly force wasn't always required: 'The noyses of our drumms of our shrill Trumpets and great Ordinance terrefyes them so they startle at the Report of them, howsoever far from the reach of daunger.'[15]

The Native Americans, on the other hand, didn't always run scared. Usually, they listened carefully. There are even records of what look like listening posts, shaped like parabolas, put up in fields by the Chesapeake people – perhaps to catch the sound of colonists stealing their corn. Further south, William Strachey discovered the Powhatan had communication networks the equal of the African talking drums discussed in Chapter 2. Up to fifty sentinels would take turns to holler every half-hour, every other sentinel being required to answer back.[16]

Clearly, the various Native American tribes could use sound well, yet in the end they were simply outgunned. And their languages, their songs, their close relationship with the natural sounds around them: all this, if not destroyed, was pushed to the edges of American life by the newer, European sounds of those who defeated them.

Is this an essential feature of all colonialism in the seventeenth century and ever since – this need not just to rule over other lands, but to silence, or at least tame, the disturbingly alien noises found there? What, for instance, was happening some 200 years later, and in another part of the world entirely?

Like America, Australia was no empty wilderness when Europeans first arrived there. Yet for many of the new settlers and explorers, the native inhabitants were viewed as just another bewildering part of wild nature. When the Englishman John Oxley was exploring the eastern river systems in 1817–18, the alien surroundings clearly perturbed him:

> It is impossible to imagine a more desolate region; and
> the uncertainty we are in, whilst traversing it, of finding
> water, adds to the melancholy feelings which the silence and
> solitude of such wastes is calculated to inspire.[17]

The only noises that punctured this melancholy silence were the winds howling, branches crashing to the ground, and 'native dogs': 'Their howlings are incessant, day as well as night,' Oxley commented.[18] For this Englishman, the Australian interior was barren, menacing and corroding – quite unlike the ordered, busy sounds and rhythms of Yorkshire where he had been born, or of the Royal Navy in which he had served for many years. Ten thousand miles away in Europe, Romantics were fashionably enraptured with the sublime quality of nature; here in the outback, John Oxley's response was more like 'horror and depression'.[19] Nor was he alone in feeling that way. Other explorers recoiled at a land that was either silent as the grave or violently disturbed by the lonely howl of wild dogs or birds screeching like 'wild spirits'.

As for the Aborigines, explorers' accounts described them as either being silent, which only went to show their unnerving stealth and perhaps their treachery, or, conversely, as savages who 'yabbered', emitted 'shouts and grunts', 'dreadful shrieks', 'hideous shouts' or 'loud and inharmonious cries'.[20] These are descriptions that echo almost word for word the colonists' language about Native Americans, and in portraying Australia's native sounds as animal-like or demonic, the same principle was being advertised: these people were simply barbaric, so they forfeited 'any moral claim to the land'.[21]

What this new land needed, it seemed to these nineteenth-century explorers, was to be filled with the more familiar, comforting sounds of home. One day, surely, Oxley wrote, it would be prime farming territory covered with grazing animals – 'bleating sheep and lowing herds' – while the 'lone and trackless' wastes would be filled with the pleasing 'hum and dust of trade'.[22] In the meantime, as in America, a strong colonial presence had to be felt. One way of achieving this was through an artillery of sound. Gunfire, for instance, was always good for keeping indigenous people in line: 'On the discharge of a double barrel … they seemed much terrified, and soon after retired,' Major T. L. Mitchell wrote in 1839.[23] And if guns proved ineffective at intimidation, there was other acoustic ammunition: rockets, bugles, speaking trumpets and gongs.[24] Usually, the best approach to unsettling sound – or even silence – was some kind of terrific, overwhelming counterblast.

When it came to sound, the colonists and the colonised had more in common than they ever cared to admit. Both Puritans and Native Americans, for example, thought of sound as having a great invisible force behind it. Take thunder: was it the voice of God or of some animal spirit? Neither answer was any more rational than the other. Yet to justify itself, colonialism needed to cast indigenous people as utterly different – as wild, savage, irrational. And sound was complicit in this dubious process. In America and Australia, it suited the new arrivals to regard what they heard there as primitive and threatening. Complex languages and varied musical traditions were lumped together and levelled down into some amorphous, primordial 'orality', then contrasted unfavourably with the essentially literate settlers. And just as written laws and proclamations gave the colonies some sense of legitimacy and permanence, so too did the ritualised beating of drums, firing of guns, blowing of trumpets and ringing of bells. It was an object lesson in the power politics of sound at the dawn of the modern age – something that applied not only to the relationship between colonisers and colonised, but also, as we shall learn, to the relationship between rich and poor.

17

Shutting In

In the eighteenth century, Scotland's capital Edinburgh was looked upon as one of the most striking cities in Europe. It was also one of the worst places in which to live: overcrowded, squalid and noisy.

Captain Edward Topham, an English army officer stationed in Edinburgh in the 1770s, would frequently walk along the High Street as it sloped up to the castle, and marvel at the extraordinary architecture all around him – especially the soaring granite-grey tenements on either side of the street: 'The style of building here is much like the French: the houses, however, in general are higher, as some rise to twelve, and one in particular to thirteen storeys in height,' he wrote in a letter in 1775.[1] In between these tall, lean tenements, there were countless narrow dark lanes or alleys, called 'wynds' or 'closes', some of which would plunge dramatically down the steep slopes on either side of the High Street. The houses would also often extend downwards at the back, creating extra hidden storeys, or cellars 'like caves slotted into the hillside', as one historian has described them.[2] As the population grew, buildings were extended over the street or sideways over the closes. The result, in Robert Louis Stevenson's words, was that 'the complication of its passages and holes' made Edinburgh 'for all the world like a rabbit warren'.[3] It also made for one of the noisiest soundscapes of any city centre in the eighteenth century.

The city's problem was that it was still hard to spread outwards. For centuries its people had huddled close to the castle for protection. After the Battle of Culloden and the defeat of the Jacobite Rebellion in 1746, the threat of war had gone. But Edinburgh was still awkwardly contained by its old fortified city walls as well as by its natural setting. Every one of its inhabitants – rich, poor or middling – had little choice but to squeeze ever closer together. 'I make no manner of doubt that the High Street of Edinburgh is inhabited by a greater number of persons than any street in Europe,' Captain Topham observed. By the early nineteenth century, there would be a dramatic change: Edinburgh's wealthier families would have their own area of elegance, privacy and quiet, having turned their backs – and shut their doors – against the noise and filth of their less fortunate fellow citizens. But until then, among the tenements and staircases and alleyways of Old Edinburgh – for rich and poor, young and old, respectable and dissolute – living side by side was the order of the day.

So what I want to ask is this: what was the soundtrack to their lives? How did these extraordinarily jumbled-up living conditions shape the sounds of Edinburgh's homes and streets? And what does their inhabitants' attitude to noise reveal about attitudes to privacy in the eighteenth century?

One tenement on the north side of Edinburgh's High Street offers some useful clues. Called 'Gladstone's Land',* this building has been authentically restored by the National Trust for Scotland and can be visited today. You enter it through a ground-floor shop, and indeed many tenements like Gladstone's Land would have had similar stores under the arcade formed by the first floor above, and facing on to the street. At the back of the shop there is a stone spiral stair-case, which takes you to the other floors. As Captain Topham wrote:

> The buildings are divided, by extremely thick partition walls, into large houses, which are here called lands, and each storey of a land is called a house. Every land has a common staircase … This staircase must always be dirty, and is in general very dark and narrow … As each house is occupied by a family, a land, being so large, contains many families …[4]

At Gladstone's Land there are six storeys, so in Topham's day at least six separate families were living here at any one time. In London the most fashionable floor would nearly always have been the first, with poorer tenants living higher up.[5] Topham, however, tells us that in Edinburgh 'the higher houses are pos-sessed by the genteeler people',[6] although clearly there was lots of variation. The original owners of this particular tenement back in the seventeenth century, Thomas and Bessie Gled-stanes, chose for themselves the third floor: a quietly elegant, comfortable space, with a painted ceiling. By Topham's time, it was the first floor that was being rented out to the wealthiest family, a merchant perhaps. It's certainly the grandest of the apartments, with a hall, two chambers and a kitchen. In one

---

* The name 'Gladstone's Land' is a little confusing: properly speaking, in Scotland the word 'tenement' refers to the land underneath, while the building itself is the 'land'. 'Gladstone' comes from the name of the family to which it first belonged, a wealthy merchant family called the Gledstanes.

of the rooms the walls are wood-panelled, which would have offered extra soundproofing. If you stand on the first floor, it's hard to think of Gladstone's Land as anything other than a desirable and peaceful town house, which in a sense it was.

But while the middle floors were occupied by the relatively rich and were comfortably fitted out, the upper floors and those below were filled by tenants of more modest means: a minister of the church in one, a shopkeeper in another, a town-hall official in a third. Another tenement nearby had a dowager countess on the second floor, a fishmonger on the ground floor, milliners and dressmakers on the upper floors, and tailors and tradesmen in the attics and basements. These other homes would have been plainer, more cramped, and, especially for those on the ground floor and basements, much closer to all the noise coming from the shop and the street at the front.

Then there was that common staircase at the back. Whatever the social distinctions between the different tenants, they all shared this staircase and had to pass each other going up and down every day. While Gladstone's Land is a relatively modest-sized building, many others had eleven, twelve or more storeys, which is why these common staircases have been described as 'upright streets', always full of human traffic.[7] There were no separate servants' quarters in the building either, so it's quite possible that anyone working for one of the other wealthier tenants would have had to rest or sleep on the staircase if they needed to be at the beck and call of their master. Even back inside, privacy was a problem. In the first-floor kitchen we find a bed attached to the wall that would have been folded out each evening, perhaps for a child or a servant to use. Cooking and sleeping, it seems, had to co-exist.

There was, then, an extraordinary social structure behind the façades of these Edinburgh tenements, with people of dramatically different status jumbled up together and required to cope with all their different lifestyles, different domestic routines and, no doubt, different attitudes to the noise they made.

But the greatest jolt to the senses would have come whenever the occupants ventured outside, into the bustling street. Before even descending the stairs and stepping into the daylight, the smarter tenants, especially the women, would probably have put on their feet some 'pattens', a kind of metal overshoe that lifted their feet and their hemlines a few extra inches off the ground. Making their way along a close and out into the street, the reason for wearing these pattens would have been quickly apparent: residents often had to walk through mud, rubbish and liquid waste, including raw sewage, which lay festering on the ground.

The first layer of noise out here, then, would have been the chinking and clattering of pattens on the cobbles as people went about their business.[8] But there were lots of other layers of sound in the High Street, since it was a maelstrom of activity. As Captain Topham wrote:

> They suffer a weekly market to be held, in which stalls are
> erected nearly the whole length of it, and make a confusion
> almost impossible to be conceived … the herb women, who
> are in no country either the most peaceable or the most
> cleanly beings upon earth, throw about the roots, stalks, &c.,
> of the bad vegetables.[9]

Even when there wasn't the weekly market there were the luckenbooths – stalls set up in the middle of the street, selling cloth, or some of the other new goods flooding into Edinburgh and other European towns from the colonies: coffee, tea, drinking chocolate, tobacco. The stallholders, along with a throng of hawkers, peddlers and ballad singers, would have been bawling loudly about their goods, competing to attract customers. Other people would gather to chat. Indeed, Captain Topham was struck by how the Scots were more like the

French than the English, in being vivacious, frank and gar-
rulous, even when it was members of the opposite sex talking together: 'they do not sit in sullen silence, looking on
the ground, biting their nails, and at a loss what to do with
themselves … they address each other at first sight'.[10] That
is if they could even hear each other against the tremendous
racket of the iron-rimmed wheels of passing carts and carriages
rattling on cobblestones, the whip-cracking of animal drovers,
and various sheep, cows and pigs being taken to slaughter, or
even the background rumble and clatter of tanners, smiths and
brewers working in the other streets nearby.[11]

As darkness fell, the amount of noise in Edinburgh's streets
would have diminished dramatically. At eight o'clock, the
shops under the arcades of the tenements closed. At ten, as
the city authorities had decreed, the 'great bell' was rung 'for
keeping good order within this burgh and eschewing trouble
and night walking'.[12] The bell was also the nightly signal for
people to hurl the contents of their chamber pots and other
household waste out of the window on to the street below –
with a warning shout of 'Gardey-loo!' – a colloquial version of
the French phrase *gardez l'eau*. For anyone below hoping not
to be spattered, the only hope was to shout up to a window
'Haud yer haun!' – hold your hand. Or, of course, not to be
out on the streets in the first place.[13]

Indeed, this was the appointed hour when, not just in Edinburgh but across the whole country, all decent folk would
be 'shutting in'. It was, as Amanda Vickery has described,
'a fateful moment that set in motion a ritual ceremony of
fortification': the decisive sounds of doors everywhere being
'locked with an integral lock, padlocks, internal bolts, iron
bars and wrought-iron chains'.[14] By the mid eighteenth century there was no longer a formal curfew in most towns and
cities, but early bedtimes were a sign of decency and wandering the streets at night was grounds for suspicion: it was widely

assumed that only streetwalkers and housebreakers would be lurking outside.[15] Silence equalled good order. And soon, in Edinburgh, for Captain Topham all was well:

> At eleven o'clock all is quiet … not so much as a watchman
> to disturb the general repose. Now and then at a late – or
> rather an early – hour of the morning, you hear a little party
> at the taverns amusing themselves by breaking the bottles
> and glasses; but this is all in good humour, and what the
> constable has no business with.[16]

This cheery account needs to be treated cautiously. For even at this hour not all *was* quiet: not everyone was home or in good humour, and there was plenty going on that would almost certainly have interested the constable. Indeed, at night, as the city settled down, its soundscape moved from its 'low-fi' fog of daytime noise into a kind a crystal-clear 'hi-fi'. Every rustle and footstep and human cry now stood out in stark relief, hinting at danger or mischief.[17]

In among the tenements and narrow alleyways of Edinburgh, the whole idea of 'shutting in' turns out to be a bit misleading. For one thing, there were plenty of people lurking here at night who had no front door to speak of in the first place. Take the story of John Macdonald, a servant who had originally arrived in the city as a homeless child along with his older sister and three young brothers in the 1740s. They had all been orphaned when the Jacobite Rebellion was crushed, and had walked all the way from Inverness-shire to look for work or beg. They ended up going down one of these closes to find the door to one of the tenements' back staircases:

> … we lay in the stairs; for about Edinburgh, as in Paris and
> Madrid, many large families live upon one staircase. They
> shut their own door, but the street-door is always open.[18]

In the communal staircase they lay, sheltering from the cold as best they could, but certainly not resting peacefully:

> There was an opinion at that time very prevalent amongst us poor children, of whom, after the Rebellion there were a great many, that the doctors came at night to find poor children asleep, and put sticking-plasters to their mouth, that they might be dissected; and indeed I believe it very true ... So when we passed the night in a stair or at a door, one slept and the other kept watch.[19]

For the poor and homeless, the night-time soundscape of Edinburgh was full of terror. And if Robert Louis Stevenson's description of these hidden corners was accurate, they were probably right to keep their ears as well as their eyes wide open:

> You go under dark arches and down dark stairs and alleys ... and the pavements are encumbered with loiterers ... skulking jail birds; unkempt, bare-foot children; big-mouthed, robust women ... a few supervising constables and a dismal sprinkling of mutineers and broken men from higher ranks in society, with some mark of better days upon them ...[20]

The lawyer and diarist James Boswell was not quite a broken man, but he would certainly be among those prowling the streets at night, flushed with claret, and punctuating the calm with the muffled sounds of his own revelry and debauchery:

> Wednesday, 6 March 1767. I was so much intoxicated that instead of going home I went to a low house in one of the alleys in Edinburgh where I knew a common girl lodged, and like a brute as I was I lay all night with her ...

> Wednesday, 28 August 1776. I was a good deal intoxicated, ranged the streets, and having met with a comely, fresh-looking girl, madly ventured to lie with her on the north brae of the Castle Hill ...

Thursday, 27 February 1777. I got into an uncommonly
cordial frame, and drank greatly too much. I unhappily
went to the street, picked up a big fat whore, and lay with
her upon a stone hewing in a mason's shade just by David
Hume's house ...[21]

Just how noisy Boswell was during these escapades we can
only imagine. But the Scottish courts were certainly full of
complaints about sleepness nights caused by drunken fights in
the streets or prostitutes plying their trade. Indeed, the number of complaints about nocturnal noise was rising steadily all
across Europe. In cities and towns everywhere, people were
packed together more tightly than ever, working longer hours,
needing their rest. This, of course, made disturbance more and
more likely – and each disturbance all the more exasperating.

The kind of extreme antagonism that might erupt in
crowded, disputatious conditions, when most people wanted
sleep but not everyone could get it, can be seen in the extraordinary, murderous events which unfolded in the 1730s at a
busy printer's shop in the Rue Saint-Séverin in Paris. The
young apprentices who worked there had to sleep on the
premises, in small, filthy lean-to sheds in the courtyard. Their
master and his mistress slept several floors up, in the main
building. It was obvious who had to rise first in the morning:
the apprentices. But after a hard day's work these unfortunate
young men had no guarantee of sleep. Their mistress kept
cats, which, as cats tend to do, gathered at night, screeching
and howling – in this case right outside where the apprentices
lay.

Eventually the apprentices snapped, and they decided to get
their own back. One of them, the best feline mimic, climbed
up to the roof and started to howl outside the window of their
master and mistress's bedroom. After a few nights of this, the
master ordered them to get rid of the cats quietly – provided,

of course, that they didn't harm his wife's favourite grey cat. Alas, things got out of hand. The apprentices unleashed a general massacre of cats in and around the printer's – including the mistress's own: some were bludgeoned to death; others maimed, then put on mock-trial and hanged in the courtyard among much laughter and jeering; yet others were burned on bonfires – an old French tradition, usually enacted on Mardi Gras – as crowds gathered around to hear the pitiful howling of the cats as they died.[22]

Clearly, patience was wearing thin. And the Great Cat Massacre of the Rue Saint-Séverin could easily be read as a bad omen. In eighteenth-century France the aristocracy and the bourgeoisie were increasingly nervous about simmering revolutionary feelings from below. Even in British cities, overcrowding, and the rising arc of urban noise that came with it, brought a new sense of unease. The diaries of middle-class Londoners, for instance, rail against the din of 'servants, workmen, the poor, hawkers, drinkers and men who fought each other with cudgels', and William Hogarth's famous engraving *The Enraged Musician* (1741) captured perfectly the irritation of creative souls who sought peace and quiet – and instead found only itinerant musicians and bawling street sellers stationed outside their windows.[23] In the Rue Saint-Séverin, of course, it had been the young and poor who had suffered the effects of noise. Yet it was the better-off who felt most strongly that the poor were invariably the problem – and that a bit of distance from them was just what was needed. Naturally enough, it was the rich, too, who were in a position to do something about it. Quite simply, they decided to move.

In Edinburgh the opportunities for creating oases of quiet for the middle classes were pretty thin on the ground in and around the High Street itself. A few new town houses were put up at the far end of a close, away from the worst of the noise. Some houses in the Lawnmarket were redesigned to create a bit more privacy and quiet – usually by ensuring that each home was two rooms deep, not just one. More glazing, more wood-panelling, tapestries, plaster ceilings: all of these helped deaden and insulate noises from the outside or from other rooms.[24] But this was just tinkering around the edges. The decisive change came with plans unveiled in the 1750s for a whole New Town, which, as the plans put it, would be for 'people of certain rank and fortune only'. Construction progressed slowly: money was tight, and several bridges were needed to link it with the High Street area. But some fifty years later the magnificent Georgian architectural scheme was complete, and almost every wealthy Edinburgh citizen had deserted the Old Town.[25]

The New Town was beautiful: tidy, regular, elegant and ordered. It reflected the Enlightenment impulse to create a 'new civil society', and responded perfectly to that eighteenth-century cult of 'sensibility', which favoured refinement, decency and delicacy over the vulgar and earthy.[26] But the New Town wasn't just geographically distant from the Old Town. It represented, in its architecture, a whole new attitude to privacy. As Topham wrote, 'The greatest part of the New Town is built after the manner of the English, and the houses are what they call here "houses to themselves".'[27] The old model of shared tenements with communal staircases was abandoned, and with it went that old promiscuous intermingling of social ranks. Each family now had a self-contained house of several floors and their own front door.

Every now and then these New Town residents might leave their threshold, cross one of the bridges and return to the Old Town on business or to shop. But when they did so, it would be as observers rather than participants in its old cramped, noisy soundscape. They were no longer struggling to live night and day in the midst of it all, but in all likelihood seeing the Old Town as strangely exotic. They would also notice, perhaps, that it was increasingly squalid, as their old tenements were filled by 'lesser ranks, then by poorer people, inevitably to become multi-occupied by the destitute', as the architectural historian Charles McKean puts it.[28] If, just before nightfall, they re-crossed the bridge, walked confidently along the well-lit, empty pavement of their own elegant street and shut their own front door securely behind them, they could wallow in the calm, sedate quiet of their new fortress home. This, at last, was their haven. As Emily Cockayne has written in *Hubbub*, it was the kind of space where sounds could be controlled to a greater degree than ever before.[29]

Though, of course, nothing was ever quite as simple as that. Safely cocooned from noise indoors, some residents 'would have become more attuned to it outdoors, and therefore more likely to moan about it'.[30] The quieter the background, the more that sudden and unwanted noises would have felt like a violation. Certainly, the need to complain about noise whenever its brash vulgarity threatened to encroach upon quiet middle-class enclaves would show no sign of diminishing in the century to come. Nor was the New Town home itself quite the private space its owners might have imagined when they first moved in. Indeed, its hushed rooms and corridors turned out to be the perfect setting for eavesdropping – and through that, the public unveiling of intimate secrets and sexual scandal.

18

Master and Servant

In 1796 the wealthy Scottish landowner John Lamont stepped through the front door of his new house in Edinburgh's New Town for the first time. He must have been a deliriously happy man. Number 7 Charlotte Square had only just been completed. It was perhaps the finest house on the most expensive side of the grandest square in the whole of Scotland's capital. Having no doubt lingered for a moment to enjoy the elegant façade designed by Robert Adam, Lamont would have now found himself in the middle of a generously sized lobby, from which a central staircase could take him to three more floors above or down to a basement. This self-contained four-storey house had a dining room, a large and splendid drawing room, a parlour, a master bedroom on the ground floor and another six generously sized bedrooms plus a nursery on the various upper storeys – in other words, plenty of space for himself, his wife, Helen, and his five children, John, Amelia, Norman, Georgina and Helen Elizabeth. This was a house that allowed John Lamont plenty of opportunity to show off his family's social standing and at the same time let them all enjoy a comfortable, quiet, private refuge from the bustling, prying, cacophonous world of neighbours and strangers outside.

But was it a private refuge? In the eighteenth and early nineteenth centuries a house like this was never completely private – or quiet:

> [I] heard a noise in the parlour like rattling of the fire-
> irons upon the fender. I went towards the door betwixt
> the bedroom and the parlour and heard rustling and other
> noise within it as if persons together – I looked through the
> keyhole and saw my mistress and Lieutenant Saunders lying
> upon the floor of the parlour he was above her ...[1]

These are the words of a young Scottish nursemaid, Sally Mackie, who worked in a house not far from the one in Charlotte Square. Her testimony reminds us that in Edinburgh and elsewhere, even a front door of your own was no barrier to surveillance. In the late eighteenth century 'the family' was a capacious category. It might include not just parents and children, but elderly relatives, unmarried sisters and paying lodgers.[2] And, of course, it would almost certainly include servants. Families such as the Lamonts would probably have had five or six of them under their roof. These were people who didn't just live in their master and mistress's house; often, they would be only a knock on the drawing room door or a footstep or two away. So while a good middle-class home might provide security, retreat, rest, warmth, food and the basis of a perfectly happy family life, it was also a place where everything might also be turned upside down in an instant by any servant who had heard too much. It was a place of eavesdropping, whispers, gossip and scandal – a place where domestic dramas were frequently played out in sound.

Every effort was made to avoid conflict and create privacy, of course. The interior layout of the Lamonts' house at Number 7 Charlotte Square shows how middle-class homes now tried right from the start to steer clear of the tightly packed communal living that had been a feature of the Old Town tenements. For a start, there was the typical division between upstairs and downstairs. Cooking was always smelly and, with all the hissing, beating, chopping, and clattering of copper pans that went into meal preparation, incredibly noisy too. So here, as usual, the kitchen, along with the wine cellar and scullery, was located in the basement where work would least disturb the rest of the household whatever the time of day or night.

As for the servants themselves, their bedrooms were also down in the basement; at Number 7 there was a cook's bedroom, two other bedrooms, plus a servants' hall, which had a recess with enough space for one more small bed. It's possible that other servants used the small bedrooms right at the top of the house, but that wouldn't have been ideal. In his book *The Conduct of Servants in Great Families*, published in 1720, Thomas Seaton had suggested that servants should not live in attics or garrets 'for all offensive things fall, rather than rise, and their noise by stirring is troublesome'.[3] In Seaton's mind, servants were always a hair's breadth away from 'quarrelling among themselves'. This meant they forever risked making every decent home 'a Scene of Confusion, a Place of Tumult and Noise' unless kept firmly in their place, which was – spatially as well as socially – 'underneath'.[4] Yet they could never be *too* far away. The man appointed by King Charles II to organise the rebuilding of London after the Great Fire spelled out the correct balance to be struck in designing a home: staff had to be far enough apart from the family's entertaining rooms to avoid disruption but near enough for them to hear 'the least ringing or call'.[5]

Just as important as the distinction between upstairs and downstairs was that between what we might call front stage and backstage, or between 'public' and 'private' parts of the house. So, on the first floor at Number 7, the domestic layout was typical: the large, elegant drawing room where visitors would be entertained ran across the entire width of the front of the house, overlooking Charlotte Square; and, through a very short passage with cupboards on either side, the smaller, less formal parlour faced the back – where the Lamont family and perhaps the closest friends among their visitors would have spent most of their time, reading, doing needlework and taking tea. Although the most misanthropic husbands railed against 'the Din of Visiting day', a woman's desire to receive visitors – and to show them what fine taste she had – was a fashionable feature of town life in the late eighteenth century. And the more visitors one had – in effect, the more one was 'on show' – the more necessary it became to create a completely private space like this within the home, a place for withdrawal and solitude.[6]

Just how much solitude, though? Architectural designs tried to prescribe exactly how close servants should be: where once they had bedded down in the drawing room or outside their master's door, or simply in a cot at his feet, they were now supposed to have their own quarters in the attic or basement. But there is plenty of evidence in church and court records that only the most senior among a household's servants could expect their own bed in their own room; many others would still have slept all over the house, in temporary beds in passages or in the corner of one of the public rooms.[7] This degree of proximity to their masters and mistresses was still an essential feature of their job, for they needed to be on hand at a moment's notice. And being on hand meant being within earshot. How would a servant ever know that more wine was required or another log needed to be added to the fire unless they did a certain amount of eavesdropping?[8] They had to be quiet workers, yes, but they also had to be good listeners. Indeed, in some London houses young apprentices might be used like guard dogs, very deliberately placed right next to the front door so that the slightest noise from outside would wake them.[9]

So, while families wanted to let down their guard, servants had to prick up their ears. In these circumstances, mutual trust was everything. Yet, as anthropologists have shown us, in most human societies trust and privacy have always been uncomfortable bedfellows. In the 1970s, for instance, Gillian Feeley-Harnik studied the Sakalava people of Madagascar. She found that they spent most of their time outside their houses in the company of others. Fences were disapproved of; doors were taboo. In Saklava society, Feeley-Harnik concluded, secrecy and separation indicated 'at best a lack of generosity'. Indeed, for people to choose to conceal themselves and keep quiet in a private space suggested, surely, that they were up to no good. And if there was a risk of them betraying the community, it seemed perfectly legitimate for everyone around to keep their ears to the ground; eavesdropping, in other words, was part of their instinct for survival.[10]

Not just theirs, but ours, too. For as the linguist John Locke reminds us, we have always been encouraged to eavesdrop for the good of the community. Across Europe in the Middle Ages, for example, court records were full of eavesdropping cases, simply because the Church encouraged people to feel a moral obligation to monitor the behaviour of others. We are not so very different today, with our 'Neighbourhood Watch' schemes, or listening out for anything that sounds like domestic violence coming from next door. As for the eighteenth century, when houses had more walls and doors than ever before, putting themselves in the way of easy surveillance, the families behind them must have imagined they were free to 'be themselves' – to talk more candidly, behave more naturally, be more conspiratorial, more flirtatious. Our observer of Edinburgh social life, Captain Edward Topham, hinted as much in his description of the private suppers that were then fashionable in the New Town:

> … when the restraints of ceremony are banished … you see people as they really are … During the supper, which continues some time, the Scotch ladies drink more wine than an English woman could well bear … and probably in some measure it may enliven their natural vivacity.[11]

It's not hard to imagine that as the evening progressed at one of these bibulous soirées the servants standing the other side of the door would have become more and more inquisitive about what was going on inside the parlour. When people were 'being themselves', the prospect of catching a whiff of improper behaviour must have been tempting. Who knew what sort of titbits might be overheard – and could later be passed on as valuable gossip?

By all accounts the Lamonts of Number 7 Charlotte Square behaved themselves impeccably, but in the country at large, servants often ended up as witnesses, hauled before the courts to testify against their master or mistress in cases of alleged adultery. Take the sad tale of John Burt, a navy captain, and his young wife, Harriet, which came to light in Canterbury in 1778. John's accusation was simple – that Harriet:

> … was and is of a lustful and lewd disposition; and during the absence of her said husband in the Mediterranean, contracted improper intimacies with divers strange men, who used to visit at unreasonable hours: that John Barlow, lieutenant of

his Majesty's tenth regiment of dragoons … came frequently
to visit …[12]

John Burt was away at sea a lot, so the crucial evidence of his
wife's alleged infidelity came from two servants, including the
cook, Susanna Huckstepp. According to her, when the master
was away, strangers used to visit Mrs Burt:

> … and they used to come towards nine o'clock in the
> evening, and stay till about twelve o'clock … one gentleman,
> called Captain Barlow… was very frequent in his visits to the
> said Mrs Burt, in the day-time, as well as in the evening.

The problem, Susanna Huckstepp told the court, was that she:

> … frequently heard strange noises in an evening in the back
> parlour, when … Mrs Burt and Captain Barlow, or any other
> gentlemen, have been there together alone: and … suspected
> that … Mrs Burt was then carrying on an adulterous
> connection.

One evening, though, it wasn't noise but a suspicious silence
that proved too much to resist. Susanna Huckstepp knew full
well that Captain Barlow was alone with her mistress. 'Not
hearing any stir or noise in the parlour', she said:

> [I] went out of the kitchen which opened into the yard, to
> see if [I] could look through the … parlour-window; and
> in order to know what Mrs Burt and Captain Barlow were
> about, did accordingly look through the window … and
> there were candles therein …

And then Susanna saw it all:

> Mrs Burt lying on her back on two chairs … her petticoats
> and shift were up, and her legs were wide extended; and she
> also plainly saw her thighs; and Captain Barlow was then
> lying upon her, and his breeches were down …

There was no doubt in Susanna's mind as to what she had seen: she had stayed at the window, she said, 'for about eight or nine minutes'.

The evidence from the second servant was less explicit but equally damning. Mary Stacey worked not in the Burt family home but at Mrs Burt's sister's house. It was here, she told the court, that Mrs Burt would come to meet a Mr Ayerst or a Mr Luttrell in an upstairs dining room, where the window shutters were closed and they could stay alone in the dark for an hour or so. Mary described what happened on a couple of occasions when she went up the stairs, as she said, 'on purpose to listen':

> [I] heard once the ... sopha crack; and it seemed to [me] as if Mr Ayerst and Mrs Burt were then on the ... sopha, and in the act of adultery ... and ... one evening ... when [I] happened to be in the drawing-room [I was] over-hearing Mrs Burt, who was alone with Mr Luttrell, tell him that he did not use her kindly, and did not come to her so often as he used to do, though he promised: and she also said to him ... 'you use me as you please' ... or she used words to that very effect ...

The evidence of the two servants carried the case and John Burt got his divorce – as did other men in other towns after countless servants had told countless courts about hearing beds 'shake' in a bedroom or the 'noise' of copulation coming from the parlour while the master of the house had been away.[13]

What is striking in all these stories is that while lawyers usually required an act of adultery to be seen, it was very often hearing something that had first raised the servants' suspicions and guided them towards the misdeed. They were ear-witnesses first, eyewitnesses second.

No wonder that the masters or mistresses of a house were

anxious, complaining of servants being 'feverishly inquisitive', or reading in *Town and Country Magazine* that 'servants are domestic spies'.[14] Even a century later, the best published advice they had was far from reassuring:

> Your servants listen at your doors ... and repeat your spiteful
> speeches in the kitchen ... and understand every sarcasm,
> every innuendo ... They understand your sulky silence,
> your studied and over-acted politeness. The most polished
> form your hate and anger can take is as transparent to these
> household spies as if you threw knives at each other.[15]

In truth, eavesdropping rarely gave servants much power. Yes, they might gossip away a person's reputation in a matter of seconds; they might even make a little money from black-mail. But usually if they ever heard anything scandalous they would find themselves in an impossible situation. After all, the typical servant was a teenage girl answerable to an older, male householder. She would have to take sides, and there was the risk she would be thrown out of her job if she chose unwisely; telling all might bring the whole household down in ruins. Sometimes a discreet silence was the best option; sometimes she had little choice but to hold her tongue. In an earlier period, men would have used a scold's bridle to physically silence any woman who risked spreading harmful informa-tion about their private affairs; by the late eighteenth century they would be encouraging their servants to read instructional manuals that told them to 'Avoid tale-bearing, for that is a vice of a pernicious nature, and generally turns out to the disad-vantage of those who practise it'.[16] As a category of speech, gossip was systematically ridiculed and denigrated.

Middle-class families such as the Lamonts had one other option to help dispel any nagging feeling that servants were constantly eavesdropping. Their house in Charlotte Square was equipped with the very latest technology: a set of bell-pulls in each room, linked by wires to a line of bells at the foot of the stairs leading into the basement. There was even a primitive kind of telephone system, allowing family members to speak from one room to another via a set of concealed pipes and tubes. Thanks to these, servants could be summoned from the kitchen or wine cellar to the first-floor drawing room or even a second-floor bedroom at the flick of a wrist or with the briefest of words. In the dining room there was another domestic device popular in the late eighteenth century: a dumb waiter – indeed, a pair of dumb waiters at each end of the table. These would have allowed the empty dishes to be whisked back and forth between the kitchen in the basement and the guests on the ground floor without any servant having to be present at all.[17] The advantages of this were obvious. As James Boswell wrote in May 1775, while dining with his wife's 'dearest friend' the use of a dumb waiter had allowed them to flirt 'with unreserved freedom'. It meant, he explained, that they had had 'nothing to fear'.[18]

We tend to think of communication technology as bringing us ever closer together, yet here in Charlotte Square, and in the thousands of other homes now equipped with bell-pulls and dumb-waiters, technology was all about pushing people further away. Servants no longer had to hover by doors or in corridors: they could be isolated from the family in the furthest corners of the house and still be summoned when required. For people such as the Lamonts it was privacy at last. They had gained a real chance to create a family life free from being overheard, even in the family home.

But privacy came at a price. If the anthropologists are right, something was lost when we started conducting more of our lives behind walls and closed doors. Hearing one another, knowing that we ourselves are being heard, can be a source of comfort rather than anxiety, and of trust rather than fear – as one Londoner recognised when she recorded the sounds coming through the walls of her own flat on a typical day in 1899:

> I hear the tenant overhead, Mr A., getting up for his day's work ... His wife, who does a little dressmaking when she can get it from her neighbours, was up late last night (I heard her sewing machine ...), so he does not disturb her ... But before he has done, I hear a child cry; then the sound of a sleepy voice, Mrs A., recommending a sip of tea and a crust for the baby ... the crying ceases, and I hear his steps as he goes downstairs. At eight o'clock there is a good deal of scraping and raking ... This means that my neighbour, Mrs B. ... is raking out and cleaning her stove. Then the door is opened ... and a conversation is ... carried on by two female voices. Among other topics, is the favourite one of Mrs A.'s laziness in the morning ...[19]

The noise of people moving about nearby, half-heard conversations, even gossip: these provide a background sound that reassures us that we are not alone. Conversely, to be out of earshot is sometimes to be out of mind – isolated, neglected, misunderstood.

19

Slavery and Rebellion

For servants in eighteenth-century Edinburgh, the kind of social soundproofing they experienced made for a more rigid class divide, but it was hardly a matter of life and death. Among African American slaves, however, the struggle to make noise and be heard was crucial to their very sense of being, and among white plantation owners the power to enforce silence would be something they would fight to defend in the most brutal fashion. We hear this struggle over sound at its most ferocious – but we also hear the first notes of a new, distinctly American musical culture being born – in moments of outright rebellion.

On the morning of Sunday, 9 September 1739, near the Stono River on the outskirts of Charleston, South Carolina, one of the largest and most violent slave revolts in American history took place. A group of men rose up in armed revolt against local plantation owners. Within days more than fifty people were dead, the rebellion had been put down, and vengeance exacted. The slaves themselves left behind no clues about why and how they rebelled. The only version of events we have is from the whites who crushed them. Even so, what evidence there is reveals one curious feature of the whole affair: the role played by the beating of drums. Sound, it seems, had a pivotal role at Stono – for both the rebels and their opponents. For one group, it was a means of expressing a shared African heritage; for the other, an unnerving sign of secret communication, which had to be stopped at all costs. What, then, was going on? Why was the beating of drums so important to the events unfolding near the Stono River that day?

The following bloodcurdling descriptions of the Stono Rebellion, written not long after it took place, supply a few basic facts:

… their Negroes had made an Insurrection, which began first at Stonoe … where they had forced a large Store, furnished themselves with Arms and Ammunition, killed all the Family on that Plantation, and divers other white People, burning and destroying all that came in their Way …

Colonel William Stephens, 13 September 1739.[1]

… they had marched to Stono Bridge where they had Murthered two Storekeepers[,] Cut their Heads off and Set them on the Stairs[,] Robbed the stores of what they wanted and went on killing what Men, Women, and Children they met, Burning Houses and Committing other Outrages …

General James Edward Oglethorpe, 17 September 1739.[2]

… in less than Twenty four hours they murthered in their

way there between Twenty & Thirty white People & Burnt
Severall houses before they were overtaken, tho' now most of
the Gang are already taken or Cut to Pieces ...

Robert Pringle, Merchant, Charleston,
26 September 1739.[3]

... and pursuing after them, within two Days kill'd twenty
odd more, some hang'd, and some Gibbeted alive ...

'A Letter from South Carolina', *Boston Weekly
News-Letter*, 1 November 1739.[4]

These fragments allude to the rebellion's beginnings: the Saturday night, when a handful of slaves working in the rice plantations attacked Hutchenson's stores, a small wooden building which stocked farming tools, general household supplies, alcohol – and, of course, weapons. The accounts then suddenly mention plantation houses being burned down, the slaughter of their owners, and, before we know it, the rebellion's disastrous finale, when the rebels, having been dispersed, were hunted down one at a time in the fields and forests and rivers and creeks of the humid, mosquito-infested South Carolina Low Country, to face summary justice.[5]

But why, by daybreak Sunday, had an isolated attack on a humble shop turned into something apparently larger, more threatening? By then the slaves who had been in Hutchenson's stores and plundered several houses were walking along the Pon Pon Road, a dusty track heading south from Charleston towards Savannah. There were already between forty and sixty of them, and other slaves were joining them all the time. It looked as if a small riot might be turning into a more widespread uprising, and what created this impression was not just the violence they had wrought but the way they were walking and the sounds they were making as they went on their way:

... they calling out Liberty, marched on with Colours

displayed, and two Drums beating, pursuing all the white
people they met with … They increased every minute by
new Negroes coming to them, so that they were above Sixty,
some say a hundred, on which they halted in a field, and
set to dancing, Singing and beating Drums, to draw more
Negroes to them, thinking they were now victorious over
the whole Province, having marched ten miles and burnt all
before them …[6]

These men weren't exactly part of a well-drilled army, but
neither were they just walking: they were *marching* – with
colours displayed. And they had drums beating – drums which
announced their progression and called out to others in the
huts and fields nearby.[7]

Ironically, it was this noise that proved the rebels' undo-
ing, since it allowed local militiamen to find them easily. But
what is striking is that even though the rebellion was crushed
quickly, the plantation owners remained nervous for months,
if not years. And what unsettled them most was the rebels'
use of drums.

Most plantation owners had always kept a watchful eye on
what their slaves got up to, even inside their own huts. A 1688
account by the British doctor Hans Sloane, from the colony
of Jamaica, tells of slaves living in huts very much like the
ones on the South Carolina plantations with only the barest
essentials: a 'Mat to lie on, a Pot of Earth to boil their Victuals
in' and a hollowed-out calabash or two to use 'for Cups and
Spoons'.[8] Their grinding daily routine would begin with an
assault upon their ears: 'They are rais'd to work so soon as the
day is light, or sometimes before by the sound of a Conche-
Shell, and their Overseers noise, or in better Plantations by
a Bell.'[9] And they would then be expected to work in the
fields without complaint or chatter. Indeed, slaveholders often
bragged about their ability to impose silence on their workers.

Sometimes, of course, they would order them to sing, on the basis that it just might improve their efficiency – as Frederick Douglass described in his autobiography:

> Slaves were expected to sing as well as to work. A silent slave was not liked, either by masters or overseers. 'Make a noise there, make a noise there' were words usually addressed to slaves when they were silent. The remark was often made that slaves were the most contented and happy labourers in the world, and their singing was referred to – and proof – of this alleged fact.

Silent or singing, it was the owner or overseer who was largely in control of the plantation's soundscape.

Yet even in Jamaica slaves had been able, every now and then, to make sounds of their own choosing. Hans Sloane gives us a glimpse of the music they played when left alone:

> The *Negros* … although hard wrought, will at nights, or on Feast days Dance and Sing … They have several sorts of Instruments in imitation of Lutes, made of small Gourds fitted with Necks, strung with Horse hairs, or the peeled stalks of climbing Plants … These instruments are sometimes made of hollow'd Timber covered with Parchment or other Skin wetted, having a Bow for its Neck, the Strings ty'd longer or shorter, as they would alter their sounds … They have likewise in the Dances Rattles ty'd to their Legs and Wrists, and in their Hands, with which they make a noise, keeping time with one who makes a sound answering it on the mouth of an empty Gourd or Jar with his Hand. Their Dances consist in great activity and strength of Body, and keeping time …[10]

Sloane clearly regarded the slaves' music-making as rather primitive, but he seemed to detect additional layers of meaning lurking just beneath the surface. He noticed, for instance, that the dances required 'strength of Body, and keeping time'. He also noticed the curious absence of certain instruments:

> They formerly in their Festivals were allowed the use of Trumpets after their Fashion, and Drums made of a piece of hollow Tree ... But making use of these in their Wars at home in *Africa*, it was thought too much inciting them to Rebellion, and so they were prohibited by the Customs of the Island.[11]

Some forty years before the Stono Rebellion, then, the colonists of Jamaica had decided that drums or trumpets were dangerous. By 1699 they had also been banned in Barbados. In 1711 St Kitts followed suit.[12] In every case, it seems, what had counted against them was their reputation in Africa.

It had been Christian missionaries, settled in Africa since the sixteenth century in places such as Ghana, Nigeria, Angola and the Congo, who had first raised the alarm over drumming. The 'Hellish' drum, one of them wrote, was 'commonly made use of at unlawful Feasts and Merry-makings'.[13] It's quite possible the missionaries were also witnessing the 'talking drums' we met in Chapter 2. They certainly knew that horn and drum musicians, especially in West Africa, were associated with kings and their courts, and therefore suspected that there was some kind of military signalling contained within their pounding rhythms.

All of this seemed threatening enough to white colonists living on, say, the Gold Coast or in Angola. But, of course, the very Africans who played these drums and horns were now being sold to slave traders and then exported to the West Indies or America. Although only a very few actual musical instruments crossed the Atlantic with them,[14] it wasn't the hardware so much as the slave's musical traditions that worried the plantation owners of Jamaica. The sugar barons were keen to ensure that nothing could be heard on their own land that sounded in the slightest bit like secret signalling or an incitement to battle.

This same anxiety later spread northwards to South Carolina. Plantation owners there included many who had first made their fortunes in the sugar fields of the West Indies; they also made use of slaves from the Caribbean as well as from West Africa. They were never terribly interested in the African heritage of these slaves, but they had heard reports of the drumming. And in the Stono River area they might also have known that many of their slaves were originally from Angola or the Congo, where there had been large-scale warfare in the eighteenth century. There would be experienced soldiers in their midst.[15] This was probably why the Stono River rebels banging their drums and singing in time set the plantation owners' nerves jangling in 1739: it would have hinted at a wider plot unfolding.

In fact, there is no evidence of an organised conspiracy at Stono River. Most likely, the slaves were angry at being forced to clear snake- and insect-infested ditches on a Sunday, their one rest day. They might also have been incited to march south by an offer of freedom issued by Spanish authorities in Florida eager to disrupt British interests. But what mattered was what was in the fevered heads of the plantation owners. Even after the rising was crushed, they couldn't shake off an uneasy feeling that collusion was still all around. When a rumour spread among them, 'it gained a kind of corporeal shape, ghosts becoming substantial',[16] and it made perfect sense to South Carolina lawmakers to assume that all such fears might be allayed if slaves faced certain new prohibitions, namely 'using or keeping of drums, horns, or other loud instruments which may call together, or give sign or notice to one another of their wicked designs or purposes'.[17] As Richard Rath points out, once this new code was introduced in 1740, 'mentions of slaves playing drums virtually disappear from colonial records in South Carolina and Georgia'.[18]

Yet the plantation owners had been so focused on the dangers of drums and trumpets that they overlooked all the other ways in which slaves sustained their African traditions. The music Hans Sloane witnessed in Jamaica in 1688 had been performed on stringed instruments, and after 1740 it was fiddles that replaced drums in the slave huts of South Carolina and Georgia. Charleston high society approved, since fiddles echoed respectable European traditions of violin playing. No doubt, too, many slaves, especially second-generation slaves, 'learned the music that Europeans wished them to play'.[19] But these fiddle players were also drawing on older West African traditions of stringed-instrument playing. They also improvised, using small sticks to strike the fiddle softly and create a distinctive three- or four-note rhythm. This was not just an echo of African traditions; it was also a subtle reinvention of those very drumbeats banned by the white authorities.[20] And where fiddles or sticks weren't available, they would use spoons or their own hands. So although by the twentieth century drumming itself had long ceased to be a central part of African American culture in the South, quite specific drum-like rhythms could still be heard in the hand-clapping patterns that accompanied, say, Sea Island spirituals or prison work songs.

The 'call-and-response' style of this kind of music points to another aspect of slave culture which saw African and European influences mixing together to create something distinctly new in the American soundscape. This was the sound of speech – and especially of religious speech. The call-and-response is a vocal pattern we associate most closely with the polyphonic rhythms of central Africa, but many of the slaves in the Stono River Rebellion were Christians, having already been converted in Africa by Portuguese or Spanish missionaries. When the celebrated preacher George Whitefield came to

Georgia and South Carolina in 1738, huge crowds would turn out in the rice fields to hear his outdoor sermons. At first, the slaves who came to hear Whitefield had no buildings of their own in which to worship. They would hold services in 'invisible churches' in the woods, muffling their voices by praying into upturned pots and dampened blankets to avoid detection by their owners.[21] Eventually, rough-and-ready churches were created, where slaves were free to express themselves in an unrestrained way, to find a moment of emotional release in an otherwise un-free existence.

By the 1850s, when a curious newspaper journalist called Frederick Law Olmsted – a man used to the order and reverence of Northern services – stumbled across these Southern African American churches, he was struck by how utterly alien yet undeniably exciting they were. In New Orleans he listened to what he called the 'incomprehensible' but somehow 'beautiful' language of a sermon, and then watched the response of the congregation:

> … an old negro near me … trembled, his teeth chattered, and his face, at intervals, was convulsed. He soon began to respond aloud to the sentiments of the preacher …: 'Oh, yes!' 'That's it, that's it!' 'Yes, yes – glory – yes!' and similar expressions could be heard from all parts of the house whenever the speaker's voice was unusually solemn, or his language and manner eloquent or excited. Sometimes the outcries and responses were not confined to ejaculations of this kind, but shouts, and groans, terrific shrieks, and indescribable expressions of ecstasy – of pleasure or agony – and even stamping, jumping, and clapping of hands, were added. The tumult often resembled that of an excited political meeting; and I was surprised to find my own muscles all stretched, as if ready for a struggle – my face glowing, and my feet stamping – having been infected unconsciously,

as men often are, with instinctive bodily sympathy with the excitement of the crowd.[22]

For Olmsted it was all a 'jumbled cacophony': his ears weren't attuned enough to detect the richness of real meaning and the blend of traditions in these sounds. But he recognised, at least, that the call-and-response, the easy mixing of music and speech, the playful language in this African American service offered a more collective, more involving style of worship than was on offer in the buttoned-up white churches back home. This was more than a matter of religion, however: the slave's style of speaking saturated the whole landscape of the American South, with its distinctive cadences and tones. It gave its speech a pervasive musical quality.[23]

From the perspective of the twenty-first century it's not hard to look back and trace a rich through-line of aural history: from the drumming of West Africa and the impromptu concerts of those slaves in Jamaica in 1688, through the noise and confusion of the Stono Rebellion, the fiddle music, hand-clapping, preaching and singing, all the way to the jive, ragtime, blues, jazz, gospel and hip-hop, which have become firmly established as part of mainstream American culture. Indeed, those sounds of Southern slavery have shaped culture globally. They have certainly become part of what Caryl Phillips calls an 'Atlantic Sound', linking South Carolina to West Africa via the shipping routes of the Middle Passage.[24] For in the bars, cafés and nightclubs, not just of New York or London, but of a coastal city such as Ghana's capital, Accra – a place closely associated with the old slave trade but also now one of Africa's most culturally open centres of trade and tourism – it's easy to be surrounded by rhythms and beats which we think of as essentially American. Walking through its bustling, pulsing heart, we hear that the music has, so to speak, returned home.

Of course, in Accra, just as in America in the eighteenth

and nineteenth centuries, music is something that is never static for long. Accra is a cosmopolitan city – of Ghanaians and Nigerians, descendants of the old Ashanti kingdom, of Dutch and German and Portuguese missionaries and British colonists, followers of Islam and Catholicism and Pentecostalism: a city of three million or so people from countless traditions and ethnicities.[25] So it's full of the noises of improvisation and creative adaptation, from African jazz and highlife, through to Africa's own take on hip-hop – 'hiplife' – and even the extraordinary honking horn music performed at funerals by groups of taxi drivers. In the history of sound, the circle is complete though unending.

The African slaves who worked the rice fields at Stono River in the 1730s had few freedoms. But one they *did* have was the freedom to adapt the sounds their owners suppressed into new forms that took on a life of their own. When, two centuries later, Ralph Ellison wrote his book *Invisible Man*, he showed the difficulties faced by African Americans in becoming socially visible. But he also showed how having a voice can make up for invisibility – how African Americans' growing and distinctive audibility might play a vital role in the whole Civil Rights movement. Democracy has always been, in part, an aural struggle – a struggle to be heard in ways of one's own choosing. For African Americans, it was a long, slow struggle to turn sounds originally regarded as 'inappropriate' into an accepted – indeed, celebrated – part of modern, shared cultural life.

20

Revolution and War

Revolution and war are unlikely to be quiet affairs. For those caught in the thick of the upheaval and violence, the experience might even be defined by noise more than anything else; what they remember, even years afterwards, is the visceral shock or disorientation of overwhelming and unceasing din, the terror or the foreboding at the danger it foretells, the sheer auditory theatre of it all:

> … one huge sea of heads covered the whole Place [Dauphine] and thousands and tens of thousands were wrapt in confusion, noise and violence.[1]

> … mutterings of the awful strife … began to be heard. Soon the din began. The voices of an hundred big mouthed guns began to vomit.[2]

The first of these two eyewitness accounts is of Paris on the brink of revolution in August 1788; the second is from an infantryman in the midst of an American Civil War battle three-quarters of a century later. Both prove – if proof were needed – that when political and social differences erupt into fighting, an onslaught on the senses will quickly follow.

Yet the soundtrack of the French Revolution wasn't always that of a frenzied mob, nor was the American Civil War all about the noise of unceasing bombardment. In both revolution and war, conflict was made of a richer, more complex tapestry of human sounds. Protest songs and rabble-rousing speeches, the ability to be silent and listen out for the invisible movements of your enemy: attending to these, as well as to the obvious din, allows us to experience the conflicts differently. It reveals how often sound itself had a major role in shaping the way the two events unfolded.

In the case of the French Revolution, one place to begin is with the sounds drifting playfully through the air, across the parks and along the streets of Paris in the 1780s. For what Parisians heard then, when out and about in the city, would perhaps have given them a keen sense of what was in the air politically.

The liveliest place to spend your leisure time if you were a citizen of Paris in the final days of the monarchy would probably have been in the gardens of the Palais-Royal, which had just been opened to the public but remained closed to the police. You would have been able to stroll among the crowds, soaking up the atmosphere of street performers, boulevard theatres, sideshows, shops, pickpockets and prostitutes. You would have walked past people selling pamphlets and newspapers offering you a profusion of reading material, much of it stuffed with the latest scurrilous gossip about the royal family or the king's ministers – gossip you might chew over later with your drinking companions at one of the busy cafés. On the Pont-Neuf, and along nearby boulevards, you would also have heard strolling ballad-mongers as they hawked satirical and bawdy songs. And when you stopped at a café for a drink, you would most likely have overheard some of these songs being performed – songs of love and seduction, but also of politics: the extravagance of the court, the king's impotence, the queen's sexual favours. You might even have joined in: the tune would be well known, and the rough-and-ready lyrics easy enough to remember.[3]

Paris wasn't yet in open revolt, but the gardens of the Palais-Royal were an extraordinary cultural melting pot. As Simon Schama puts it, they had succeeded in bringing a 'raw and Rabelaisian popular culture right into the heart of royal and aristocratic Paris'.[4] Here, if nowhere else, you couldn't miss the sound of French political life shifting: the indiscriminate mingling of different classes, the loosening of rank and hierarchy – the lack of deference. When revolution finally erupted in 1789 and 1790, it would drink lustily from this brew of noisy irreverence.

Music and singing, especially, turned out to be important to the Revolution. Thousands of songs were written and performed in cafés, theatres and public squares, the most popular evolving month by month in response to political events or the mood of the people. One, 'The Carmagnole', heaped scorn on the queen – 'Madame Veto' – and her aristocratic supporters:

Madame Veto has promised,
Madame Veto has promised
To cut everyone's throat in Paris,
To cut everyone's throat in Paris,
But she failed to do this,
Thanks to our cannons.
Let us dance the Carmagnole,
Long live the sound,

Long live the sound,
Let us dance the Carmagnole,
Long live the sound of the cannons.[5]

Even more popular in 1790 was 'Ah! Ça ira', with its message of hope rather than anger. 'It'll be fine,' went the song: everything 'shall succeed'. Performing 'Ah! Ça ira' became a symbolic show of faith in the Revolution, especially if it involved trying to out-sing royalists, who now had songs of their own. But the more people were singing 'Ah! Ça ira', the more its lyrics were improvised – and the darker they became in tone. Soon the Sans-culottes were trilling about hanging aristocrats: 'If we don't hang them, we'll break them,' they roared; 'if we don't break them, we'll burn them.' If you want to trace the changing moods of the French people during the Revolution – from optimism and magnanimity to frustration and anger – then following the evolution of their protest songs is as good a way as any.

Whatever form they took, songs were invariably brilliant at rousing enthusiasm and creating a sense of collective solidarity. They were, however, only one part of the soundscape of political mobilisation in revolutionary Paris, for there was also a thriving culture of public oratory. By 1789, coming to the Palais-Royal on a Sunday to mill around with thousands of other people was the best way to witness some spectacular speech-making. The sheer drama of it all struck one English visitor as he sat in a café observing what was going on around him:

… expectant crowds listening … to certain orators who
from chairs or tables harangue their audience. The eagerness
with which they are heard and the thunder of applause they
receive for every sentiment of more than common hardiness
or violence against the present government cannot easily be
imagined.[6]

French orators such as the young and precociously talented Camille Desmoulins had read their Cicero and Tacitus and Livy. Desmoulins regarded himself as the inheritor of this Roman tradition. His style, though, was not to speak to his peers, offering a nod to the crowd as and when required, but to appeal directly to the crowd itself, to draw upon its emotions:

> Listen, listen to Paris and Lyon, Rouen and Bordeaux, Calais and Marseille. From one end of the country to the other, the same universal cry is heard ... everyone wants to be free.[7]

This style of oration, as Simon Schama says, was full of 'breast-beating' and 'sob-provoking declamation'. The crowd played its part, too, with shouts of 'bravo' before carrying their hero off 'amidst a great shouting and cheering'.[8]

The French Revolution, then, was forged not so much by people reading pamphlets and filling their heads with radical ideas in the privacy of their own homes, as by people gathered together in public spaces, listening to the revolutionaries themselves. It was a revolution forged by audiences. Men like Desmoulins often preferred enlisting support through public speeches than through writing. It wasn't that he felt he could rely purely on emotions and forget about ideas: these weren't crowds inflamed beyond reason – at least not yet. It was more that when Desmoulins talked, he knew that what he said was invested with personality, made human. A pamphlet could be suppressed or amended, reinterpreted, even misinterpreted. But his own voice was unmediated, apparently spontaneous in expression, indivisible from the man; in short, it announced his qualities as a personality as well as a thinker. Being a little emotional sometimes, rather than always restrained, was no longer a sign of being primitive or vulgar; it was a sign of sensitivity, of having a feeling heart.[9] As for those in the crowd, they were far more than passive bystanders: in 1789 they felt more like actors in 'a brilliantly lit Historical Moment'.[10]

In the sound of the Paris crowds we don't just detect a revolutionary mood drifting through the air, we hear a new kind of politics emerging. The rigid and artificial world of ritual and decorum at the royal court at Versailles, where politics had consisted of tightly controlled spectacles and guarded utterances designed to preserve the mystique of absolute power: all this was giving way to something more expressive, raucous, even earthy. The politics of passionate speech on the street was more organic, of course. Feelings were liable to spin out of control.[11] And if this new, noisy kind of politics was excitingly democratic, it was also in danger of provoking other, more ominous sounds: the dying forces of the Old Regime meeting like with like – trying to overpower the sounds of revolution with their own armoury of noise. As audiences were replaced by rioters, the Paris city authorities mustered their own support through that recognised signal for times of peril, the sounding of the tocsin (alarm bell) – reinforced, when necessary, by cannon shots and the beating of drums. Meanwhile, mounted soldiers from outside the city moved in on behalf of the king and queen and unleashed their own volleys of gunfire over the crowds. The point of outright revolution hurled together the noises of uprising and the noises of authority to create one thunderous cacophony. The uproar would pass. But it was at moments like this that the sounds of revolution were barely distinguishable from the din of all-out war.

Any soldier who fought in the American Civil War of the 1860s, of course, would tell you that war *proper* is different: altogether louder, more relentless, the overwhelming nature of the sound a portent of mass destruction. In October 1862 a reporter for the *Charleston Mercury* recalled for his readers the Battle of Corinth in Mississippi, where Confederate forces were driven back by heavy artillery: 'The air was filled with the missiles of death, and the earth trembled under the confused noise of battle.'[12] Clearly, with vast quantities of ordnance being let off, rifles firing, drums beating, the tread of cavalry, the wheels of gun carriages grinding away and soldiers screaming and yelling, being in the midst of any battle would have been, as one soldier said, like finding yourself in a 'deafening … seething hell' of din.[13] Indeed, part of the noise was quite deliberate – a technique for intimidation, as it had been in warfare for centuries. In the American Civil War there was, for instance, the 'rebel yell' of the Confederates, fabled for its ability to terrorise the enemy with its high-pitched ululating, whooping and howling.[14] Yet of course Union troops shouted too, and yelling only really worked when backed up with firepower.

Besides, the war was marked more often by noiselessness than by noise. For soldiers, there were long periods spent waiting, listening – time enough, indeed, for the 'still quiet' of the South's rural landscape to impress itself on many town-bred Northerners, and for Southerners in turn to become anxious at the sounds of conflict rupturing the precious tranquillity of their home front.[15] Sometimes, alas, the silence was sickening – a marker of too many dead in the field, or of that dead calm which precedes a storm. And then there were the times when commanders imposed silence on their troops as a means of moving by stealth – as one Union soldier recalled: 'With such noiseless caution was the retreat conducted, that the falling back of our skirmishers was unknown to the enemy.'[16]

Against this background, any sudden or violent noises, whenever they did arise, took on extra significance. Far from being a meaningless cacophony, they were valuable clues about the progress of fighting. One Confederate soldier, for instance, described what he heard at a crucial moment in the first major battle of the war in July 1861, the Battle of Bull Run in Virginia:

A fire of musketry and artillery was beginning to develop

on the left ... The roar of the young battle now swelled in
volume. There came crashes of musketry which told that
whole brigades were coming in, and the fire of the guns
increased ... About 11.30 am ... There came a further access
of fire both of musketry and artillery ... No one who heard
it could doubt its import. No messages from the left were
needed now. All paused for a moment and listened ...[17]

So soldiers listened carefully, using sound as a source of military intelligence. This usually served them well. They would
know, for instance, not to move forward until they had heard a
concerted volley of their own guns, or to regroup if a direction
of attack seemed to shift.

But sound could also be ambiguous: when soldiers relied
too much on their sense of hearing, they could also be deceived
– sometimes with disastrous consequences. This happened in
February 1862, when the Confederates managed to launch
a surprise assault on Union forces surrounding them at Fort
Donelson in Tennessee. The Union troops had the advantage, but one morning around dawn the Confederates managed to attack – and it was some six hours before their enemy's
key commander, Ulysses S. Grant, who was away from the
field discussing tactics, was told about it and could arrange
for reinforcements. On the face of it, it's hard to explain how
he didn't hear the battle raging for himself: he was only four
miles away, and by all accounts the fighting was monumentally
loud. 'The firing was very heavy and continuous, being musketry and artillery mixed,' one source records; 'Terrific can-
nonade ... An almost incessant discharge of musketry ... The
fire upon our lines continued with unabated fury.'[18]

So why hadn't Grant heard any of this? One likely reason
is that he was in a kind of acoustic shadow, where the sound
of battle was blocked from reaching his ears. Between him
and the battlefield there was an intervening forest, and the

trees and ridges probably baffled the uproar of the Confederate assault. There had also been a heavy snowstorm brought in by a 'driving north wind' just a day or so earlier, and the snow would have absorbed other sounds as it lay on the ground. To make matters worse, Grant was upwind of the battle. Altogether, he stood almost no chance of catching what was going on in the field of battle itself. Fortunately for the Union troops in the end, Grant was able to turn matters around and the Confederates were eventually forced back into their fortifications.[19] Even so, it was touch and go. A commander's reliance on sound as an early-warning system had nearly proved fatal. It is one more example of how the noise of warfare has never been solely about sheer volume, the exposure of soldiers to indiscriminate din. The noise of warfare has also been about the need for soldiers to train their ears and listen carefully. Above all, it has been about sound's consistent refusal to serve just one master in any human dispute.

The American soldiers of the Civil War, just like the Bourbons of Versailles with their displays of pomp and power, and the New England colonists we met in Chapter 16, had sought to harness the power of sound to mesmerise or over-awe or terrify – or even simply to send vital information flying through the air. But in doing so they had placed their faith in a capricious force. True, the sound of bells and drums and guns could be deployed in the service of the powerful, as could the simple ability to impose silence on others, and in the eighteenth and nineteenth centuries many sounds remained the markers of authority. But sound also slipped its chains. Everyone, it turned out, could find a way of making noise of one sort or another. Even those who were otherwise powerless – the colonised, servants, slaves, revolutionaries, soldiers under siege – could find ways of breaking the silence imposed on them, of using sounds of their own, sometimes in the service of their own liberation.

That sound has a certain democratic quality to it was understood well enough even at the time. Why else, after all, would plantation owners in Jamaica or the American South have sought to suppress the music of their slaves? Because they knew the sounds they made when they had the chance had a power all of their own. Understanding this was a sign that the whole notion of an aural culture, where speech and music are rich in meaning, was by no means a disappearing feature of humanity's 'primitive' or pre-literate past. It was ever-present, happily co-existing with a world of print and reading and writing and imagery, and, it seems, constantly evolving. In the Middle Ages, outside the world of scholarship, sound was still commonly invested with mysterious qualities, either divine or demonic. It was understood to have a tangible force. By the nineteenth century it was less mysterious, but a force nonetheless: no longer a force of nature or of the spirits to which people had to submit passively, but something they could take hold of and shape quite precisely in order to manage people's feelings and convey information. As we stood on the brink of the modern age, with its rush of industry, machinery, new communication networks, scientific advance and dazzling visual spectacle, sound – and listening carefully to sound – was going to be just as important as ever to the way people lived their lives.

V

The Rise of
the Machines

21

The Conquering Engines: Industrial Revolution

A few miles north-west of Boston, Massachusetts, just beyond the outer reaches of the city's sprawling suburbs, is the small, neat, historic town of Concord. By heading south from Concord's centre along a reasonably quiet main road, it's possible, in less than ten minutes, to reach an idyllic, near-circular lake surrounded by pine trees, which also just happens to be one of the sacred places of American literature: Walden Pond. The lake is not really far enough away from American urban civilisation to count as true 'wilderness', but in 1845 it was certainly sufficiently peaceful to attract the attention of one of the United States' most talented writers, then living in Concord: the 28-year-old Henry David Thoreau. He built a simple one-room cabin on the north side of the lake and there he lived for the best part of two years, recording this extended communion with nature in what became his best-remembered work, *Walden*.

At the heart of *Walden* is Thoreau's discovery of how in this secluded spot he had the time to enjoy what he called the 'bloom' of the present moment:

> Sometimes, in a summer morning, having taken my
> accustomed bath, I sat in my sunny doorway from sunrise
> till noon, rapt in a revery, amidst the pines and hickories and
> sumachs, in undisturbed solitude and stillness, while the birds
> sang around or flitted noiseless through the house, until by
> the sun falling in at my west window, or the noise of some
> travelling wagon on the distant highway, I was reminded of
> the lapse of time.[1]

For Thoreau, peace and quiet was absolutely essential to this experience. It's what allowed him to notice the rich, subtle layering of nature's own sounds as he sat at his cabin doorstep:

> ... a whippoorwill on the ridge pole, a blue-jay screaming
> beneath the window, a hare or woodchuck under the house,
> a screech-owl or a cat-owl behind it, a flock of wild geese or a
> laughing loon on the pond, and a fox to bark in the night ...
> Sturdy pitch-pines rubbing and creaking against the shingles
> for want of room ...[2]

If he walked the few yards towards Walden Pond's shoreline, he would hear it ringing with the 'trump of bullfrogs'. If he walked into the woods, he would hear the 'sweet and melodious' lowing of 'some cow in the horizon beyond', and, on a Sunday, catch the 'faint, sweet' peal of bells from Concord's churches. As each peal drifted through the woods, Thoreau felt it acquired 'a certain vibratory hum, as if the pine needles in the horizon were the strings of a harp which it swept'.[3] Woven together, these layers created the colourful vernacular of Walden's soundscape. But Thoreau's lakeside meditation was also, at least in part, an elegy. He sensed this precious place was under threat. 'The whistle of the locomotive penetrates my woods summer and winter,' he wrote, 'sounding like the scream of a hawk sailing over some farmer's yard ... I hear the iron horse make the hills echo with his snort like thunder, shaking the earth with his feet ... and I am awakened ...'[4]

The sounds of the railway were still utterly novel. In their own way, they could strike people as being just as varied as the sounds of nature, which Thoreau already so admired. The Canadian writer R. Murray Schafer has rhapsodised brilliantly on this subject: 'the whistle, the bell, the slow chuffing of the engine at the start, accelerating suddenly as the wheels slipped, then slowing again, the sudden explosions of escaping steam, the squeaking of the wheels, the rattling of the coaches, the clatter of the tracks,' he wrote, when trying to capture the rich 'lore' of past sensory experiences in his pioneering book *The Soundscape*.[5] For Thoreau, though, a passing train meant more than just a new layer of noise. When a freight train rattled past, he suddenly felt more connected with the rest of the world. He thought of all the goods on board – where they had come from, where they were going – and his mind conjured up 'coral reefs, and Indian oceans, and tropical climes, and the extent of the globe'.[6] When he heard a train at a station he thought of farmers and townspeople starting to set their clocks by the sound of its arrival or the whistle-blast of its departure instead of by the chimes of the church bell.[7] For Thoreau, the sound of the train was the sound of a new age being born: Nature giving way to Machine.

If anything, Walden Pond escaped lightly. By the middle of the nineteenth century, railways were tearing into the countryside at a phenomenal rate across America and Europe. Africa and Asia would be next – Egypt got its first railway in 1852, Sudan soon after – each line speeding up the rate at which valuable minerals could be sucked out of the countryside and whisked away to the coast for export. In Britain, where the first line had been opened back in 1825, there was soon a network covering most of the country, apart from parts of Cornwall, Wales and northern Scotland.[8] Wherever the railway came, the movement of people and goods speeded up. It provided a heady rush of new sensations. It brought excitement. But for many contemporaries it was also a little frightening, as if the immense noise of the railway spoke of something bigger, something violently destructive coming in its wake.

Charles Dickens, who was writing *Dombey and Son* at almost exactly the same time as Thoreau was in Walden, described trains as 'conquering engines', with their 'shrill yell of exultation, roaring, rattling, tearing on'.[9] The ride from Leamington to Birmingham, which he describes in the novel, feels remorseless, even monstrous, as passengers fly through the countryside:

> … with a shriek, and a roar, and a rattle, from the town,
> burrowing among the dwellings of men and making the
> streets hum, flashing out into the meadows for a moment,
> mining in through the damp earth, booming on in darkness
> and heavy air, bursting out again into the sunny day so
> bright and wide; away, with a shriek, and a roar, and a
> rattle, through the fields, through the woods, through
> the corn, through the hay, through the chalk, through the
> mould, through the clay, through the rock …[10]

For a passenger the noise was real enough. But the railway was just one layer in a new and apparently all-encompassing din: the sound of a steadily advancing industrial revolution. In the countryside, threshing machines were replacing the sickle and scythe; in mill towns, power-looms were turning textile-making from a cottage industry into a vast, mechanised factory system; all sorts of manufacturing now involved steam engines, hydraulic presses, pistons, iron grinding against iron.[11] It was as if super-human, or rather *in*human, noise was smothering an organic soundscape that had gone undisturbed for centuries.

Mostly, the countryside heard merely the distant ripples of the Industrial Revolution. Its crucible was in cities like Manchester or Birmingham. By the middle of the nineteenth century, they were so transformed that any train passenger alighting in them for the first time would have been over-whelmed. In 1845 the Scottish folklorist and geologist Hugh Miller visited both during a tour of England. As his train snaked into Birmingham he noticed first the 'long low sub-urbs', then, in the city itself, an entirely alien soundscape:

> In no town in the world are the mechanical arts more noisy; hammer rings incessantly on anvil; there is an unending clang of metal, an unceasing clank of engines; flame rustles, water hisses, steam roars, and from time to time, hoarse and hollow over all, rises the thunder of the proofing-house ...[12]

The 'proofing-house' was where gun-makers would test their newly manufactured weapons. It would have sounded to any-one passing close by as if a platoon of soldiers were inside the building, firing volley after volley. And the noise would barely have abated. Birmingham was producing on average a new musket every minute, night and day, throughout the year.[13]

What would it have been like for the workers themselves? Wisely, no doubt, Hugh Miller declined to go inside. But we've a vivid account of the cacophony of an iron-mill from Thomas Carlyle, who had visited Birmingham in 1824:

> ... blast-furnaces were roaring like the voice of many whirlwinds all around; the fiery metal was hissing thro' its moulds, or sparkling and spitting under hammers of monstrous size, which fell like so many little earthquakes. Here they were wheeling charred coals, breaking their

ironstone, and tumbling all into their fiery pit; there they
were turning and boring cannon with a hideous shrieking
noise such as the earth could hardly parallel.[14]

Once they were operated by steam engines and lit by gas, Britain's mills and factories had little need to stop and catch breath. For those toiling away inside, the working day stretched, shifts multiplied. Most had to live as near to their workplace as possible, or even in among the machinery itself. In 1832, investigators found children sleeping regularly at one factory so that they wouldn't face the owner's wrath for being late. In other words, hundreds of thousands, perhaps millions of people were now trapped into experiencing near-continuous noise.

It wasn't long before the awful effects on their health were plain to see. Imagine, for instance, the impact if you were unlucky enough to be one of the boilermakers in Glasgow observed by Dr Thomas Barr in 1886:

In the process of boiler-making, four different classes of men are engaged – riveters, caulkers, platers, and 'holders-on'. The riveter drives in with a large hammer the red-hot iron rivets for binding the plates together; the caulker hammers with a chisel the edges of the plates so as to ensure complete tightness; the plater forms the iron plates and arranges them accurately in position; while the holder-on stands inside the boiler holding a large hammer, the head of which he presses against the inner end of a rivet … The men who work inside the boiler, such as the 'holders-on', are, of course, exposed to the loudest and most damaging sounds. Their ears are near to the rivet which is being hammered in by the riveter outside. The iron on which they stand is vibrating intensely under the blows of perhaps twenty hammers wielded by twenty powerful men. Confined by the walls of the boiler, the waves of sound are vastly intensified, and strike the tympanum with appalling force, while the vibrations from the iron

pass directly through the bodies of the men to the delicate
structures of the inner ear ... After such an experience one is
surprised that the delicate mechanism in the interior of the
ears can retain its integrity for a single day ...[15]

It's no surprise to discover that Dr Barr's medical practice
was overflowing with boilermakers, nor that 'not one of them
had normal hearing'.[16] And by the end of the nineteenth century, 'boilermakers' disease' was by no means restricted to just
that one industry. It was found regularly among shipbuild-
ers, locksmiths, iron-turners, weavers, engine drivers, railway
workers.[17] Back in Walden Pond, Massachusetts, Thoreau's
greatest worry had been the occasional passing train. Those
whose job it was to lay the tracks and stoke the boilers for
those trains would have suffered the din of Industrial Revolution in its most violent register. For them, the machine age
was, quite literally, deafening.

Was there *any* chance for workers to flee the grating noises
of industry, even for just a moment? Well, on Sundays a few
might take a short train ride into the country in pursuit of
some peace and quiet. But not everyone could afford to do so,
and few could afford to worry about the price they would pay
in poor health by staying in the city week after week, month
after month, year after year. Many simply accepted – or, rather,
were forced to accept – that noise was an inevitable accom-
paniment to their lives. According to our Scottish traveller
Hugh Miller, there were families in Birmingham who lived
right next to a foundry, surrounded by the endless noise of
hammering, yet 'slept soundly night after night'.[18] Looking
for the faintest glimmering of good news, Miller wondered if
perhaps Birmingham's rich musical culture was a by-product
of this industrial din. Its citizens, he reckoned, having lived
'in an atmosphere continually vibrating with clamour', had
absorbed a love of noise into their very souls. Now they sang

and played in concerts and festivals in order to fight factory noise with sounds of their own.[19] Even if this were true, the message was still grim. If Birmingham folk really were deliberately filling in every possible moment of quiet, then it seemed as if a dreadful kind of sonic arms-race had taken hold of society – as if for many people it was no longer noise but silence that was unnatural; as if people had now submitted themselves completely to the machine.

Sounds, as we know, don't reach our ears without changing us in some way. They shape our moods and regulate our lives. During the Industrial Revolution, submission to the sounds of the machines meant submission to their needs and rhythms. When Hugh Miller's tour of England took him to Manchester, he noticed that within a minute or so of the city's clocks striking one, the hitherto deserted streets would be full of a rushing 'human tide', like 'a Highland river in flood'. It was clear that in all the factories and warehouses nearby a shift was ending and 'the dinner hour of the labouring English' was underway. Within minutes the streets would be empty again. An hour later, the human tide would return, summoned once more by the sound of the striking clock or the shrill blast of the mill's whistle announcing a new shift.[20] In this ebb and flow, Miller could detect a new kind of humanity being forged.

The new machine-age human was captured brilliantly much later, in the 1930s, when the BBC broadcast a portrait in sound of a Sheffield steel-mill. Its producer, Geoffrey Bridson, mixed recordings inside the mill with orchestral music and chanted verse to weave together what became an aural epic. We hear men making their way to the factory gates, then clocking on, the transformation of pig iron into milled steel, and a voice-over itemising the phenomenal amount of raw materials used and outputs achieved by the end of a gruelling working day – all marshalled into one sonic onslaught. With its monumental sound, the programme certainly conveys industry's resounding roar.[21] Yet, strikingly, at no point do we actually hear the steelworkers speaking spontaneously in their own voices. At first, this seems bizarre. Indeed, the folk-singer Ewan MacColl complained that in this respect Bridson's programme was guilty of effacing the individuality of those men at the heart of the process. But of course that was precisely the point. Bridson had wanted to convey the abstract power and energy of industry, and the new social order it created.[22] This, his programme seemed to be saying, was a new order based on new rhythms; the people were cogs in the machine – nameless, faceless, *voiceless*.

Before, the rhythms of labour – tilling the field, hauling in fish, spinning cloth – had often been matched to the rhythms of the human body, the rhythms of breathing, of bending, of hands and feet moving, by people singing work-songs. These songs made the work more bearable, offering labourers a chance to sing about each other, make jokes about village characters and dabble in matchmaking. Work proceeded at a slightly different pace, one set by nature and local conditions: seasonal variations in weather, perhaps, or the life cycles of animals stirring, sleeping, breeding, hibernating. Industrialised work tore up these delicate organic relationships. Like the heroine in Thomas Hardy's *Tess of the d'Urbervilles*, workers could henceforth only stand 'mutely' before the 'penetrating hum of the thresher'. The sheer racket of it all simply made singing or chatter impossible, and the rhythms of nature no longer mattered. The demands of efficiency required workers to synchronise themselves with the machine.[23] And, of course, the machine was there to make money. Its noises and rhythms were viewed as almost sacred, representing both progress and productivity.[24] For the good of business, people just had to put up with them.

In some ways, the whole history of sound hinges on the Industrial Revolution. According to R. Murray Schafer, it was then that the world moved from 'hi-fi' sound, rich in variety, where every nuance of the natural soundscape might be appreciated, to a 'lo-fi' existence, where sounds were drowned out by the 'cacophonies of iron' and replaced everywhere with an urgent but 'flat-line' industrial drone. Here, Schafer suggests, is a possible explanation for the listlessness of modern life. He quotes the French philosopher Henri Bergson, who once posed the question: how would we know if some unseen power suddenly doubled the speed of everything happening in the universe? The answer Bergson gave was quite simple: we would notice a great loss in the richness of experience. 'Even as Bergson wrote,' Schafer concludes, 'this was happening.'[25] The opportunity for people like Henry David Thoreau to catch the 'bloom' of the present moment in a place such as Walden Pond was disappearing fast.

But did the nineteenth century *really* leave us with nothing but a 'flat-line' of machine noise? It's possible to argue that during the Industrial Revolution, owing to human ingenuity, a whole set of other cultural and social changes meant we could hear steadily *more* sounds – even a greater variety of sounds. People were retuning their ears, learning to value listening as an art, even starting to fight back against urban noise. And while industry drowned out some of nature's sounds, science and technology was busy revealing others, even stranger, which had always existed without our ever noticing.

22

The Beat of a Heart, the Tramp of a Fly

In January 1780, a sixty-year-old Edinburgh man walked slowly through the streets of his home city, wheezing and puffing rather alarmingly. As he reached Infirmary Street, just to the south of the Old Town, he turned into Surgeon's Square. By now, he was really struggling to catch his breath, but he had reached his destination and he managed to climb the last few steps leading into the jumble of buildings that made up the University of Edinburgh's medical school and teaching hospital. John Farquhar had come all this way to see Francis Home, one of the university's most respected doctors. In 1780, if anyone in Britain had a hope of finding an accurate diagnosis and decent treatment, here was as good a place to start as any. Edinburgh was just beginning to get a reputation as one of the world's great centres of medical learning, and was already ahead of anywhere else in Britain or America – perhaps, in the not-too-distant future, even a potential rival to Paris. And Dr Home was Edinburgh's Professor of Medicine. Mr Farquhar must have thought he was in the safest possible hands.

His first consultation seemed to go well. Noting that Mr Farquhar had 'a pain in the region of the liver, especially on being pressed',[1] Dr Home regarded the diagnosis as straightforward. It looked like a swelling of soft tissues through an accumulation of excess water – in other words, a common case of 'dropsy'.[2] But it was soon clear that Mr Farquhar's troubles were spreading rapidly throughout his body. By the middle of the month his belly had begun to swell, and at the beginning of March his legs had swollen too: 'he can scarcely lie horizontally in bed, and starts, for fear of suffocation when going to sleep,' Dr Home recorded. Even his various bodily functions were changing: his urine was 'of deep colour, diminished in quantity'.[3] Nevertheless, Dr Home was convinced Mr Farquhar had dropsy, so there was no reason not to give him the traditional remedy: a purgative made of cream of tartar.

Yet instead of getting better, Mr Farquhar was dead within the week. All along, his doctor had been focusing on the symptoms and had failed to diagnose the underlying cause: congestive heart failure. Mr Farquhar's legs had swollen because his heart had been unable to pump blood properly; his weariness and feeling of suffocation had come because his lungs had filled up and been blocked by blood; and his urine had darkened because his kidneys were giving way. All this had been happening deep inside Mr Farquhar's body; medically speaking, it had all been happening been out of sight – and therefore out of mind.

Dr Home was by no means incompetent, but like almost every doctor at the time, his great mistake had been not to *listen* to his patient's body. He had assumed that because something could not be seen, it also could not be heard – indeed, that there was nothing really there to be heard in the first place, other than a patient's breathing and pulse. Yet if only Dr Home and his patient had been able to meet up in Edinburgh some forty years later, their consultation might well have had an altogether happier outcome. The early years of the nineteenth century saw advances in science which were to lead to a wholesale revolution in listening. New technology would suddenly allow people to hear for the first time sounds that had always existed – but below the threshold of normal human perception. As a result of these discoveries, the spectrum of sound expanded dramatically. As well as hearing the booming din of industry at one end of the scale, people would now be able to detect the smallest, subtlest sounds imaginable at the other. And in witnessing this rather strange microcosm of noise for the first time, their understanding of how the invisible world teemed with life – as well as their relationship with fellow human beings – would be utterly transformed.

If the revolution in our understanding of the hidden workings of the body had a beginning, it was a pretty bizarre one. In the 1750s, inside the wine cellar of a hotel in Graz in Austria, the young Leopold Auenbrugger had regularly watched his father tapping wine casks to work out how full they were. A decade later, when Auenbrugger had qualified as a doctor in Vienna, he decided the same effect might be found with people. He realised that if he placed his hands on a patient's chest, tapped with one finger, then felt for the resulting vibrations – while perhaps also pressing his head against the patient's chest to listen – he could work out whether it was healthily full of air or dangerously flooded with liquid. A highly resonant response to his finger tapping meant a good clear cavity lurking beneath the surface. Dullness, on the other hand, meant trouble – a lung filling up with pus, perhaps even a tumour.[4]

This was a potentially exciting medical discovery, but there were two problems with Auenbrugger's technique. For a start, the sounds coming from deep inside the body were still very faint: even with an ear pressed right up against a chest it wasn't easy to hear them. And perhaps more seriously, in the 1760s, pressing your ear against someone's chest, even if you were a thoroughly respectable doctor, was a rather unseemly thing to do. It was a problem that certainly exercised one French physician, René Laennec, who wrote: 'the repugnance which every one must feel to apply the ear to a patient that is dirty or whose chest is bathed in perspiration, must always prevent the habitual or frequent use of this method.'[5] Even if a patient *was* clean and wholesome, there might be a further delicacy. What, for instance, if she was a woman? As Laennec went on to say: 'in the case of females, exclusively of reasons of decorum, it is impracticable over the whole space occupied by the mammae.'[6]

Clearly, what was needed was something that allowed a bit of physical distance between doctor and patient, and, after watching schoolchildren playing with long hollowed-out sticks, René Laennec reckoned he had found a solution:

> I rolled a quire of paper into a kind of cylinder and applied one end of it to the region of the heart and the other to my ear, and was not a little surprised and pleased, to find that I could thereby perceive the action of the heart in a manner much more clear and distinct than I had ever been able to do by the immediate application of the ear. From this moment, I imagined that the circumstance might furnish means for enabling us to ascertain the character, not only of the action of the heart, but of every species of sound produced by the motion of all the thoracic viscera ...[7]

Laennec had created the stethoscope, and armed with this instrument he proceeded to map in minute detail the internal soundscape of the human body. He was now able to listen to the chest of one of his suffering patients, and the moment the patient died he could open up the body on the autopsy table to see the condition of their heart and lungs. In this way, he could link certain sounds made by the body directly to certain physical changes. A gurgling, cavernous sound, with perhaps each word spoken aloud followed by a puff: tuberculosis. An absence of respiratory noise in the chest: emphysema.[8] In this new diagnostic soundscape, coughs were not just coughs. Some were crepitous rattles, others mucous and gurgling; some were dry and sonorous, others dry and sibilant. Indeed, such was Laennec's desire for precision that he went as far as defining one particular type of cough in the following terms: a 'dry crepitous rattle with large bubbles or crackling, utricular buzzing, amphoric resonance'.[9]

In linking each ailment to a specific noise, Laennec reflected that wider nineteenth-century zeal for codifying and cataloguing all manner of things, from plants and animals to the shape of human heads, in the hope of ridding the natural world of its irritating uncertainties and irregularities.[10] As it later transpired, Laennec had been far too optimistic. Something as resolutely idiosyncratic as the human body just couldn't be treated in this systematic fashion. The basic idea of using a stethoscope to listen carefully to a patient, however, had definitely pointed the way forward. It allowed the patient's body to speak for itself. It opened up to scientific scrutiny a whole new invisible world of fleshy structures and spaces and movements. The possibilities were seized upon eagerly by a new generation of experimenters, both in Laennec's Paris and at Edinburgh's fast-expanding medical school.

At first, Edinburgh's doctors were tentative. They knew it took time for their ears to become trained: they needed retuning, in order to interpret the unfamiliar and infinitely varied soundscape of the body revealed by the stethoscope. But there were enough individuals of influence determined to advocate

the new approach – among them William Cullen, a house surgeon at the Royal Infirmary, and Professor Andrew Duncan
Junior. By 1822 Duncan was reporting the results of his own
experiments with mounting excitement:

> Anyone who will take the trouble to apply the cylinder with
> tolerable attention … to the chest of half a dozen patients in
> a hospital will easily satisfy himself of the great diversity to
> be observed in the action or rhythm of the heart, and in the
> sounds caused by respiration, and as these depend upon the
> physical state of the organs, the one is a certain indication of
> the other.[11]

Yet using this cylinder in a doctor's consulting room was
proving a lot harder than it looked. The problem was one of
design. The first stethoscopes were basically just rigid pieces of
hollow wood: to get one end pressed firmly against the skin of
a patient's chest and the other pressed firmly against a doctor's
ear involved all sorts of bodily contortions and uncomfortable
postures for both parties.[12] Some stethoscopes were hinged
in the middle, which helped a little. But it was the development of a flexible tube in 1828 by physicians of the Edinburgh
Royal Infirmary that really made the difference. Henceforward
not only was it possible to range over a patient's body more
freely, but there was also the avoidance of any embarrassment.
'As it does not require the head of the stethoscopist over the
chest of the sick person … it can be used in the highest ranks
of society without offending fastidious delicacy,' wrote one
of them.[13] The doctors of the Royal Infirmary declared that
this new flexible tube also made it altogether less daunting
to examine people at the other end of the social scale, those
who were 'manifestly contagious or miserably wretched'.[14] By
the 1830s, then, there seemed to be no social reason – and
certainly no technical reason – for doctors to avoid listening
attentively to anyone.

Even now, though, stethoscopes made their way into day-to-day clinical practice only very gradually. Edinburgh had benefitted from Scotland's historic ties to France, allowing it to pick up on developments in Paris much faster than the rest of Britain. Edinburgh was also ahead of its time because in the university and the Royal Infirmary it had a well-funded and centrally organised medical service which brought patients face to face with a critical mass of expertise. London, by contrast, was hopelessly uncoordinated. Clinical practice was dominated by private practices, the medical schools were scattered and working in competition with each other, and anti-French prejudices were stronger. There was little sharing of good practice.[15]

As for America, then as now it struggled to keep up with Western Europe in its ability to treat the mass of ordinary patients. The stethoscope was introduced to the country for the first time only in 1846. Even then many of its doctors remained indifferent, largely because they were undereducated in science. In the 1770s, America had about 3,500 doctors, and of these only 400 or so had medical degrees. The situation had gradually improved by the mid nineteenth century, since more of them were by then travelling to Edinburgh or Paris for their training. After 1870, America's own medical schools, in Harvard, Rochester, Johns Hopkins, and Chicago, also finally began to teach the use of the stethoscope,[16] but that important shift from cutting-edge laboratory research to practical application had been frustratingly slow. The ability to make a reasonably accurate diagnosis on the basis of the sounds lurking inside a patient's body only became a standard feature of medical practice some fifty years after it had been first developed.

The stethoscope, though, was just one part of a much bigger nineteenth-century intellectual revolution. X-rays, electricity, radio waves, genes, the subconscious: all these discoveries seemed to suggest there were hidden and marvellous dimensions to the natural world. A teeming reality had always existed just beyond the reach of our normal senses and was now slowly, but rather excitingly, coming into view. In the realm of sound, this same dramatic unveiling of new worlds was achieved by that simple and now under-appreciated invention: what the Cambridge physicist Charles Wheatstone described in 1827 as a 'rudimentary, non-electric amplifier of faint sounds' – in other words, the microphone.[17] Wheatstone set out the basic principles of the device. In 1878 a music professor in Kentucky made it a practical reality. And in the same year, one of the British Post Office's engineers, William Preece, gave a public lecture on how the microphone would change the world:

> It will render evident to us sounds that are otherwise
> absolutely inaudible. I have heard myself the tramp of a little
> fly across a box with a tread almost as loud as that of a horse
> across a wooden bridge.[18]

The Victorian imagination was unleashed. A writer for the *Spectator*, for instance, declared that it might soon be possible 'to hear the sap rise in the tree; to hear it rushing against small obstacles to its rise, as a brook rushes against the stones in its path; to hear the bee suck honey from the flower; to hear the rush of the blood through the smallest of blood-vessels'.[19] This was not, as the historian of soundscapes R. Murray Schafer seems to suggest, a world sounding increasingly monotonous. If anything, it was sounding richer, more colourful, more three-dimensional than ever.

In listening so closely to the rumbles and gurgles of the human body or even to the tramp of a fly, Victorian science was treating sound, refreshingly, as something rational – something that could be 'read' for information. But this happened without extinguishing its age-old magical qualities.

On the contrary, the discovery that the world and everyone in it was secretly alive with sound stimulated people's sense of poetic wonder. This 'roar which lies on the other side of silence', as George Eliot famously described it in *Middlemarch* (1871–2), induced among many writers and artists what can only be described as a kind of giddy auditory vertigo, as they contemplated what a new rush of sensations might do to their delicate souls.[20] It didn't take long for them to decide. Maintaining their own sense of refinement, perhaps even protecting civilisation at large, would depend on their unique ability to distinguish good sounds from bad. Listening could no longer afford to be a casual affair. It needed to be a rigorous, disciplined skill – one, naturally, which only a select few would be capable of achieving.

23

The New Art
of Listening

One gorgeously warm and sunny August day in 1869, the Scottish-born American naturalist John Muir was wandering among the waterfalls, deep rocky valleys and ancient giant sequoias of the Yosemite National Park in the Californian Sierra Nevada. It was a place he adored. But on this particular day, instead of thoroughly enjoying himself as usual, he was thoroughly distracted, unable to settle or relax. As he explained in his diary the following day, what had troubled him was the way other people who had come to the park had been behaving:

> It seems strange that visitors to Yosemite should be so
> little influenced by its novel grandeur, as if their eyes were
> bandaged and their ears stopped. Most of those I saw
> yesterday were looking down as if wholly unconscious of

anything going on about them, while the sublime rocks
were trembling with the tones of the mighty chanting
congregation of waters gathered from all the mountains
round about, making music that might draw angels out of
heaven.[1]

The birdsong, the 'wind-music' of the trees, the 'joyous hum'
of insects, the 'glad streams singing their way to the sea':[2] the
sublime wonders of nature were all around, freely available to
be seen and heard – perhaps *especially* to be heard. For as Muir
once wrote, so rich in sounds was a park such as Yosemite that
'even the blind must enjoy these woods'. Yet most visitors –
busy catching fish, distracted by each other's company, driving
animals away with their chatter and their mules and horses –
simply weren't able to appreciate them. If only they could be
'devout, silent, open-eyed, looking and listening with love', he
complained, their lives would surely be so much richer.[3]

John Muir's call for a 'devout' and 'silent' form of listening
echoed a cry which reverberated right across the nineteenth
century: a need for the sublime power of particular sounds to
be experienced to their full effect. The pattern was set as the
century opened and the Romantics talked up the virtue of
listening closely to nature. Samuel Taylor Coleridge wrote of
'the thunders and howlings of the breaking ice' one winter's
night on the Lake of Ratzeburg, and how they had convinced
him that there were 'sounds more sublime than any sight *can*
be'.[4] In 1828 William Wordsworth even wrote a whole poem
'On the Power of Sound'.[5]

In the hands of the Romantics, and in the minds of natu-
ralists such as John Muir, this reverence for close listening
usually came across as reasonable, organic, even democratic
and liberating, but it also contained the seeds of an attitude to
sound that was darker and more authoritarian. In order for the
sublime to be appreciated, people everywhere, whether in the

countryside or the city, would have to reform their ways. And as the century progressed, it was this darker, controlling side to the art of listening that seemed to take root.

Just a few years after John Muir had written of his experience in California, in Bayreuth in Germany the foundation stones were being laid for the town's magnificent Festival Theatre opera house. When it was finished in 1876, the Bayreuth theatre established itself as one of the world's great temples of music and, in particular, as a mecca for those who worshipped the composer Richard Wagner, to whose work it was dedicated. And it really was worship in which they indulged. As Cosima Wagner put it when speaking of her husband's compositions, 'Our art – I dare say it – is religion.'[6] In the nineteenth century, many music aficionados, and not just the Wagnerites, came to regard their art in the same way. Yet, as it turned out, giving music a religious aura came at a price. Its most devoted followers would soon start creating their own dogma about how it should be heard, eventually crushing any behaviour that smacked of heresy. The new art of listening would turn out to be a rather uncomfortable mixture of wonder, politeness and brutal discrimination.

Only a century before, listening to music – even classical music – had been a surprisingly relaxed affair. This was largely because many people were not actually listening at all. Concerts were often held in restaurants. As Peter Gay has shown, even if musicians performed in the mansions and stately homes of the rich, it was often as 'an agreeable backdrop for flirting, gossiping, dining',[7] and if they ever got the chance to perform in an opera house, they would still have faced a restless auditorium. There was no firm etiquette about arriving on time or leaving quietly at the end; instead, there was a constant stream of noisy toing and froing. A handful of music-lovers, clutching their librettos, might be conscientiously following the action on stage, but most of the audience would be chatting away throughout the performance, 'as liveried servants hawked carafes of wine and oranges' all around them.[8] In the remote reaches of the upper balconies, prostitutes would also be plying their trade.

One could easily imagine the musicians themselves being in utter despair at this hubbub. Yet so used to it were they that silence unsettled them more. When Mozart performed in Vienna in 1781, he wrote to his father about how pleased he was at the screams of 'bravo' coming from the audience as he was playing. Orchestras rarely objected to their final bars being covered with shouting and applause. Indeed, usually neither performer nor composer could bear to wait until the very end of a piece before hearing of their audience's approval: they would have taken silence as a sign of rejection.[9]

Even in the eighteenth century, however, there were stirrings of change. Instead of offering merely a pleasing background entertainment, music was increasingly thought of as something capable of offering a sublime and deeply personal experience. For Charles Burney, who in the late 1770s was just beginning his epic *General History of Music*, music was – or at least *should* be – about being moved. Every listener, he wrote, had this right: 'a right to give way to his feelings, and be pleased or dissatisfied without knowledge, experience, or the fiat of critics'.[10] The critics, of course, continued to be influential in guiding wider public taste, and for many it was Beethoven who promised listeners the most sublime experience of all. The French writer George Sand, for instance, told Liszt of how she herself responded to the German composer's music: 'Beethoven makes you enter once again into the most intimate depths of the self; everything you have felt, experienced, your loves, your suffering, your dreams, all are revived by the breath of his genius and throw you into an infinite reverie.'[11]

What, though, did treating music in this reverential way mean for the 'ordinary' music listener? A clue can be found

in a painting now hanging at the Museo Revoltella in Tri-
este but which, right at the end of the nineteenth century,
became something of an international sensation, travelling
from country to country in the form of popular engravings
and copies. Lionello Balestrieri's *Beethoven* depicts a pianist
and a violinist playing in a small, sparsely furnished room to
an audience of five people.[12] None of them, it seems, is making a sound or even engaging with the other four. Two men
stare intently in the direction of the musicians – though are
looking through them rather than at them, as if they're hardly
there; another man leans against a wall on which Beethoven's
death mask hangs, hands in pockets and looking down at the
floor; a fourth sits bent forward, head in hands and eyes covered; a woman leans against one of the men, but ignores him
as she stares into space, transfixed. These five people do not
just remain silent; they are virtually unconscious of the world
about them, swept away to the spiritual realm of the infinite.
In other words, they are listening as if it were a sacred task.

'Worshipful silence' was being depicted on canvas because it
was now the ideal form of listening. To reach a state of ecstasy,
one needed to establish a direct and pure relationship with the
music, and that meant concentrating. Silence wasn't the same
as passivity. It had an intensity to it. It was *work*. The listener
was engaged in a monumental lifelong task of self-realisation
and cultural improvement – what the Germans referred to as
Bildung.[13] And this meant, of course, that anyone who shattered the silence would be guilty of destroying the purity of
the moment upon which this hard task depended. It was certainly acceptable to weep as you listened; indeed, why on earth
would you not, given how beautifully a piece of music would
evoke the depths of despair or the heights of love? But you
needed to weep quietly, to let a tear roll down your cheek
rather than wail out loud.

In almost every respect the expectation now was one of

respectful silence throughout a performance, even when it came to audiences in the largest and busiest of concert halls or theatres. Indeed, it was in public buildings such as these, in Frankfurt and Paris and London and New York, where growing numbers of people came to listen to famous music being performed by leading orchestras and star conductors, that the need to enforce silence posed the greatest problems of crowd control – and also, perhaps inevitably, created the sharpest social conflicts.

In Frankfurt in 1808, for instance, audiences were issued with a clear set of printed guidelines on the conduct henceforth expected of them at lectures and concerts:

> During literary or musical performances everyone is asked
> to refrain from speaking. Applause, too, expresses itself
> better through attentiveness than the clapping of hands.
> Signs of disapproval are not to be expected ... Dogs are not
> tolerated.[14]

Alas, it seems that in the first instance guidelines such as these had only a limited success: training people in the art of listening was evidently going to be a long, laborious process. So what if, to speed matters up, concert-goers could be encouraged to start policing themselves? This was certainly the hope of the New York Philharmonic in 1857 when it tried to insist on what it called 'musical good manners' in an attempt to tackle a continuing problem of 'inattention and heedless talking' during performances: 'The remedy ... lies with the audience itself,' it tried to explain in its annual report. 'If each little neighbourhood would take care of itself, and promptly frown down the few chance disturbers of its pleasures, perfect order would soon be secured.'[15] In both New York and elsewhere, in an effort to wear down resistance, rare examples of good behaviour were repeatedly praised. In March 1819, the *London Morning Chronicle* reported enthusiastically on the 'silence and attention' witnessed at a recent concert of the Royal Philharmonic. In 1896, a newspaper reviewer in Chicago noted with approval that a performance of Beethoven in the city had been 'received and followed with the closest attention and there was noticeably less conversation than on previous occasions'.[16]

All the effort slowly paid off. By century-end, it seemed, the ideal of decorum was taking hold. People were now more inclined to stay frozen in their seats, trying desperately not to rustle their programme, scarcely daring to breathe. And, Peter Gay suggests, it was precisely at this point – when listeners had finally suppressed their most boisterous instincts for the duration of a whole performance – that applause became dramatically wilder and demands for encores more frequent. The wall of rapturous noise which now met the climax to a piece was the sound of long-suppressed emotion exploding.[17]

This final triumph of respectful listening is built into the very fabric of a place such as Bayreuth. The main concert arena was designed to echo those ancient Greek theatres like that at Epidaurus, which we met in Chapter 6. It did away with centre aisles and boxes, and stripped away all decorative frills. Some of this was about trying to achieve the best acoustic quality, but mostly it was about ensuring that nothing could possibly interfere with the listener's feeling of rapture. This included any visual distraction whatsoever – including, rather dramatically, the orchestra itself, which, along with the conductor, was consigned by the architects to a deep pit out of the audience's sightlines. The Bayreuth concert hall was also the first opera auditorium in the world to regularly douse its house lights as a performance began: previously, most concert-goers would have remained brightly illuminated throughout. In this temple to Wagner, all that ever mattered was a rigorous, disciplined focus on 'the sounds and sights that Wagner had conjured up'.[18]

How could every single person conform to such rigorous

standards of behaviour? The answer, of course, was that they couldn't. And this, too, was part of the whole point of the new art of listening. To be able to keep still and listen was a measure of one's standing in the world: not only were you showing yourself to be sensitive to the emotions of the music; you were showing yourself to be an altogether more refined person – a person of class.

The distinctions of class were complicated, however. It wasn't always the hoi polloi in the cheap seats who failed to measure up. At the Paris opera, for instance, those who kept on talking regardless of convention were the 'powdered wigs' in the boxes – the thick-skinned aristocrats who couldn't care less for their fellow audience members, and who were especially immune to bourgeois criticism.[19] In Frankfurt, eighty-five years after those printed guidelines were issued, audience members were still being asked to refrain from leaving their seats during, or before the conclusion of, a piece, although this request was addressed not to the stalls, but to those 'in the boxes'.[20] The listening revolution represented a triumph not of aristocratic but of middle-class codes of conduct.

Naturally, there were many – and not just the bewigged aristocrats – who felt smothered by these conventions. Anyone who still nurtured the spirit of the Romantics, for instance, would have railed against the sheer stuffy artificiality of concert etiquette. Hence Charles Lamb's reaction, in the guise of Elia, to one particular evening's unwanted entertainment:

> I have sat through an Italian Opera, till, for sheer pain, and
> inexplicable anguish, I have rushed out into the noisiest
> places of the crowded streets, to solace myself with sounds,
> which I was not obliged to follow, and get rid of the torment
> of endless, fruitless, barren attention! I take refuge in the
> unpretending assemblage of honest common-life sounds.[21]

Undivided silent attention could never be fully enforced

on everyone, simply because it went against a basic human impulse: our instinctive and enjoyable need to hum along, tap our feet, drum our fingers or sway our hips whenever a piece of music enters our ears and invades our body. As James H. Johnson points out in his history of the Parisian music scene, for all those who felt the social pressure to conform and now sorely missed the chance to really enjoy the music on their own terms – happy, relaxed, expressively involved – the inevitable result was nothing other than sheer boredom.[22] And it wouldn't be long before many ordinary concert- and opera-goers sought easier entertainment in the raucous delights of the music hall, leaving those temples of classical music to become the ever more exclusive domain of the middle classes.

That was an augury for the future. For now, though, the Victorian era had succeeded in charging the very act of listening with extraordinary significance. One reason its literary culture was mesmerised by blindness was that it seemed to symbolise so perfectly how unusual powers of insight would accrue to anyone who could avoid the distractions of the visual world. In 1860, for instance, an article in the *National Review* wrote admiringly of the ancient philosophers who put out their eyes in order to concentrate their attention on the deepest meaning of things. In explaining why the central character in her epic poem *Aurora Leigh* was blind, Elizabeth Barrett Browning explained simply that 'He has to be blinded ... to be made to see.'[23] Even scientists were urging people to see beyond the surface of things. For the eminent physicist Oliver Lodge, for instance, it was vital, if human knowledge were to advance, that people began to explore far more than what was merely visible: 'Matter alone is what appeals to the senses ... [Yet] our animal senses give us no clue or indication to the wealth of existence operating in the intangible and the unseen.'[24] In Lodge's case, a deep fascination with what could be heard rather than seen led him to dabble in those fascinating Victorian pursuits of telepathy, the seance and spiritualism, It even lured him into becoming a leading figure in the Society for Psychical Research. But the possibilities of the 'unseen' were also what led him towards greater scientific understanding of electricity, magnetism and radio waves, and helped him lay some of the foundations of twentieth-century physics.

It was as if the human ear was slowly being perfected, as if its ability to discern the sublime, or even the hidden scientific truths of the world, was now more acute than ever. But, as we'll learn, having sensitive ears could be a curse as well as a gift. In a fast-expanding metropolis such as London or New York, the very reverence towards listening the Victorian era had been cultivating was to lead not so much to insight and wonder but to nervous breakdown and misery.

24

Life in the City

In the summer and early autumn of 1853, the respectable enclave of Cheyne Row in London was filled with dust and the loud, clattering sound of building work proceeding at a frantic pace. The urgent sawing and banging came from the house at Number 5, the home of the writers Thomas and Jane Carlyle. The Carlyles had builders in their attic because they had embarked on a radical alteration to their home: the creation of what they hoped would be a completely soundproof study. It was right at the top of the house, as far away from the street as possible; it had double walls, a roof lined with sound-muffling air chambers, and no windows – its only natural illumination coming from a skylight in the ceiling.

The Carlyles were desperate for their builders to be finished. Thomas, in particular, had always wanted complete peace and quiet in order to write. 'It is absolutely indispensable, for most people, especially for the like of me, so thin-skinned and so confused a being, to get into perfect seclusion of mind from time to time and to be well let alone,' he had written in a letter to his mother on 31 December 1852.[1] But ever since moving down from Scotland several years before, London had failed miserably to provide the seclusion he craved. As he laboured away at his writing desk in Cheyne Row, he had been plagued constantly by noise. There was, for instance, the incessant din of chickens, an irritation that had first erupted over a decade earlier, as Jane recorded in one of her letters:

> Our neighbours next Door have brought back with them
> from their summer excursion a live cock and hen! And the
> cock within these few days has been unfolding a rare faculty
> for crowing – God send it a speedy end.[2]

Far from disappearing, though, things only got worse, with Jane increasingly anxious about her husband's state of mind:

> That dreadful woman next door, instead of putting away
> the cock … has produced another … They are stuffed with
> ever so many hens into a small hencoop every night, and left
> out of doors the night long … and of course they crow and
> screech not only from daylight, but from midnight, and so
> near that it goes through one's head every time like a sword
> … If this goes on he will soon be in Bedlam … Carlyle swears
> he will shoot them.[3]

Actually, Thomas was too busy to set about the 'demon' fowl – mainly because he was so often fantasising about committing violence against a succession of street musicians who kept turning up on the street outside:

> 12 March 1853: … there has come a body of vagrant musical
> scamps, under my windows, with clatter-bones, guitars and

Nigger songs; which they are doing with a vigour beyond example, in hopes of a copper or two from somebody …

8 July 1853: I have been awakened … these two successive mornings at 5 o'clock; under my windows is a vile yellow Italian grinding …[4]

By now he seemed to be at breaking point. And summertime was especially difficult:

I have to keep doors and windows more or less open for air; and all men and animals take to giving voice in the bright weather. It is astonishing how many cocks, parrots, singing thrushes, dogs, dust-carts, dandy-carriages, do announce themselves thro' these open windows even in the quietest localities![5]

The almost-finished attic study held out his only hope. With this 'impenetrably "silent room",' Thomas assured his mother, he would be able to stay in the city for ever – perhaps one day, in his dotage, even withdraw to it entirely, all alone with his books and his papers in blissful peace.[6]

Alas, it didn't work out quite like that. Sitting in his 'silent' study for the first time, Thomas discovered that even though some of those irritating outside noises were now muffled or blocked by the soundproofing, others appeared to have been strangely amplified when set against the overall hush. The clever fourteen-hole ventilating system he had installed in order to avoid the necessity of windows seemed to channel into the room all sorts of sounds from the servants' quarters and living rooms below. Even then the skylight had to be opened in warm weather, rather defeating the whole point of the study's windowless design. By November 1853, in fact, Jane had decided that 'The silent room is the noisiest in the house' – 'a complete failure'. And her comment that 'Mr C' was now 'very much out of sorts', was almost certainly a diplomatic understatement. His mind was already turning once more to dark, murderous thoughts about the 'demon birds' in his neighbour's back yard.[7]

It's hard not to poke fun at Thomas Carlyle for being so absurdly 'thin-skinned' – and hard, too, not to feel a great deal of sympathy for the woman who had to put up with him all those years. Evidently, noise and frayed nerves went together. And it's sometimes difficult to tell in retrospect which was cause and which was effect – whether some of the din was more perceived than real. Visiting Thomas's silent study a few years ago, the American writer Garret Keizer decided that it was less a shrine for those seeking quiet than a warning oracle for them. 'You can imagine a motto inscribed over the door', he said: '"To be obsessed with quiet is never to be possessed of it".'[8]

It's also troubling for us now to hear that strong note of racial and class prejudice in Carlyle's condemnation of 'vagrant musical scamps' and the 'vile yellow Italian' organ grinder. But the uncomfortable truth is that his attitude was entirely typical of a growing militancy about noise among London's Victorian middle class – even among its reforming middle class. Charles Dickens, for instance, also grumbled about 'brazen performers on brazen instruments, beaters of drums, grinders of organs, bangers of banjos, clashers of cymbals, worriers of fiddles, and bellowers of ballads'.[9] Newspaper articles, meanwhile, happily engaged in full-scale racial profiling of the city's street musicians, declaring that there was 'Scarcely one Englishman, not one Scot' among them; they would, one journalist claimed, be 'nearly always a foreigner'.[10] There's no doubt that some of these musicians really were involved in a kind of extortion racket, deliberately making a nuisance of themselves outside someone's home until they were paid off. But, as the literary historian John Picker has shown, their misfortune was that they would also now come to embody the idea that aliens were overrunning England, or that refinement was struggling to keep at bay a rising tide of vulgarity. The mounting calls in Parliament for all street musicians to be silenced or somehow swept away from respectable neighbourhoods showed just how easily the issue of offensive sounds could be transformed into a rather nasty form of social discrimination, even a kind of ethnic cleansing.

The irony of all this is that street musicians irritated people such as Dickens and Thomas Carlyle, just as the faint rumblings of domestic activity below had tormented the latter in his soundproof study, simply because such noises stood out so sharply against what must, relatively speaking, have been a luxuriously *quiet* background. Elsewhere in London by contrast, especially in its bustling centre, there really was a rising arc of background sound both loud and continuous enough to shatter nerves, and it was this broader soundscape that would later prompt the medical writer Edwin Ash to declare that a new disease of the modern age, which he suggestively called 'Londonitis', was upon us.[11] The number of people and the amount and range of traffic pouring into cities like London around the turn of the century were rapidly rising. Some sounds were dissipating – by the late 1920s the motor car with its rubberised wheels had largely replaced the horse-drawn carriage with its clattering iron-clad wheels; many of the old noisy cobbles had been ripped up to be replaced by flatter, smoother road-surfaces; engineering improvements, such as the use of 'bogey' wheels, had made railways less shatteringly loud and rackety; and electrical machinery had displaced the thump and hiss of old steam-driven machinery.[12] But it would have been hard to notice such improvements among the extra noise being created by all those extra inhabitants and vehicles.

One person who was especially alarmed by this deteriorating situation was a Glasgow doctor called Dan McKenzie. In 1916 he published a book, *The City of Din: A Tirade against*

Noise. This, he makes clear, is a metaphorical city of din, not a real place, yet it drew heavily on his own experiences:

> The roar of the traffic of motor-buses, taxi-cabs, and motor-cars is of a deeper, more thunderous, and more overpowering nature than in former days, principally because the vehicles are heavier and are driven at a much greater speed.[13]

And while the clatter of iron wheels on cobbles was receding, there was a new source of irritation in its place:

> The motor horn! I often wonder why in all the world such an instrument of torture has ever been permitted to exist even for a single day! ... Horns! Surely never before in the whole raucous history of din have such fiendish contraptions split the air.[14]

This, of course, was on top of all the other stuff – the 'dirge-like' rumbling of trams, the street sellers, the 'blaring' of music.[15] And what worried Dr McKenzie most was that this unprecedented array of noise might be slowly but inexorably destroying the mental health of city-dwellers:

> The modern mind is a delicate instrument, the needle-indicator of which trembles and oscillates to the finest currents of thought and feeling. By culture and education we have acquired the sensibility of the artist or poet. And yet we continue to expose this poised and fragile instrument ...[16]

Merely by the steady daily bombardment of background city noises – noises sometimes already taken for granted – the human brain, he believed, would become first jaded, then completely worn out. The result would be an epidemic of nervous exhaustion. So the noises that had plagued Thomas Carlyle – the occasional inconvenience of a street musician or the sound of his neighbour's hens – were irritating but basically inconsequential. The bigger threat, according to

McKenzie, was the chronic low-level din experienced by hundreds of thousands of ordinary men, women and children going to school or commuting daily into city centres by train, working for long hours in shops or factories surrounded by machines and then returning home at night to overcrowded tenements overlooking restless streets.

At the close of the nineteenth century, the most densely packed place on earth – and therefore the place where more people risked nervous exhaustion through noise than anywhere else – wasn't London, however. It was the Lower East Side of Manhattan in New York. About three-quarters of a million immigrants were arriving in America every year; in some years it was over a million new arrivals. Nearly half stayed in New York.[17] Of these, the majority settled in the Lower East Side, creating in the space of just a few square miles a dazzling mosaic of nationalities and cultures – Irish, German, Polish, Italian, Jewish – and, as one history of the area puts it, powering 'the combustion engine of the city to an unprecedented heat'.[18]

Jacob Riis describes this area in his 1890 classic, *How the Other Half Lives* – the sights, smells and sounds of streets overflowing with life, as people hawk and peddle their goods, stop and chat, argue and fight and play:

> The crowds that jostle each other at the wagons and about the sidewalk shops, where a gutter plank on two ash-barrels does duty for a counter! Pushing, struggling, babbling, and shouting in foreign tongues, a veritable Babel of confusion.[19]

Here, too, New York's elevated railway snakes close by and 'noisy trains speed over the iron highway', the day begins 'at all hours' and, instead of sleeping quietly at night, the district 'carouses boisterously'.[20] Riis calls this corner of Manhattan 'Jewtown'. And if ever there was a real rather than metaphorical City of Din, this was surely it.

There is plenty of joyousness and warmth to be found in these turn-of-the-century aural snapshots of the Lower East Side. Then, as now, noise was a powerful symbol of vitality – of community, industriousness, progress. But there was a darker aspect to life here too, where endless noise exacted a heavy toll on human happiness. The tenements lining the bustling Lower East Side streets – big buildings, six, seven or eight storeys high – were described by the English novelist Arnold Bennett as places which 'seemed to sweat humanity at every window and door'. In 1895 there were some 40,000 tenements there, and together they housed a staggering 1.3 million people.[21] Jacob Riis calculated that about 330,000 people were crammed into one square mile.[22] Possibly the most crowded block in the district ran along part of Orchard Street: in 1903 roughly 2,000 people lived in just this one short stretch. If you had been a resident back then, going through the front door of one of those tenements you would be entering not so much a quiet haven but a place where families were 'stacked atop one another like so much cord wood' – a teeming, cramped world, utterly lacking in privacy, and rarely still or quiet.[23]

Jacob Riis captures the moment one young girl enters a tenement and climbs the stairs, having gone out to fetch some water. It is, he writes, 'a place where no sun-ray had ever entered':

> … up the narrow, squeaking stairs the girl with the pitcher was climbing. Up one flight of stairs, over a knot of children, half babies, pitching pennies on the landing, over wash-tubs and bedsteads that encumbered the next – house-cleaning going on in that 'flat'; that is to say, the surplus of bugs was being turned out with petroleum and a feather – up still another, past a half-open door through which came the noise of brawling and curses …[24]

The people who lived inside a tenement like the one at 97 Orchard Street (now a museum and a National Historic Landmark) – people such as the Gumpertz family, or the Confinos, or the Rogarshevskys – clearly made the best they could of their home: walls were decorated with floral paper, photographs were hung, family mementos displayed, food put on the table. But the real-estate agents and boarding-house keepers who ran the place had an economic incentive in squeezing as many tenants into it as possible, creating the perfect conditions for continuous and inescapable noise, night and day. There, and all across the Lower East Side, extra rooms were built at the back, toilets were kept outside to maximise space and, despite being dark, damp and unventilated, cellars were opened up – often in order to lodge so-called 'rent babies', young children from broken or homeless families whose parents would pay a modest rent to keep them off the streets.[25] The landlords also partitioned apartments that had once housed families in reasonable comfort – and kept partitioning them. 'Where two families had lived ten moved in … It was rent the owner was after,' Riis writes.[26]

Shrinking room size wasn't the only problem. If the partitions ran at right angles from the front and rear walls, each room might have its own window. More often the partitions ran sideways across each floor. This meant windowless interior rooms without direct light or ventilation, and people having to walk through one room to get to another. It was harder and harder to call a place like this home – and, indeed, the tenements were sometimes described as dens, barracks, hives or even rookeries. Whatever they were called, privacy was virtually impossible. In Cedar Street, Riis came across five families – twenty people of both sexes and all ages – living in one twelve-foot by twelve-foot room in which there were just two beds, a chair and a table.[27] Everywhere, makeshift beds were created from crates or rugs spread on the floor. There would be no moment of quiet, simply because, day or night, at least one person would always be stirring.[28]

In time, fear of disease prompted the civic authorities to pass laws ensuring more air and light was let into these tenements. But little was done about their noise, not least because the local economy by now depended entirely on their rooms remaining hives of industry. It was here, beyond the reach of health officers or factory inspectors, and so beyond the reach of any means of regulating hours or protecting children, that a vast army of people were at work making clothes:

> The homes of the Hebrew quarter are its workshops also … You are made fully aware of it before you have travelled the length of a single block in any of these East Side streets, by the whir of a thousand sewing machines, worked at high pressure from earliest dawn till mind and muscle give out together.[29]

In the same rooms where meals were taken and clothes were washed, perhaps a dozen people – men, women and children – might have been working, a dozen machines whirring long into the night. 'These are the economic conditions,' Riis reminds us, 'that enable my manufacturing friend to boast that New York can "beat the world" on cheap clothing.'[30]

What made clothes cheap and plentiful made sleep precious and rare. The noise of the sewing machines, the coming and going all hours, the frequent fire alarms: here, in the Lower East Side, nerves must have frayed easily. The nearby Ward's Island asylum had more than its share of inmates from the tenements. As a newspaper reporter explained in 1894, simple poverty was one factor: 'Where a woman is old or sickly, and, therefore, useless in the hard struggle for existence which constantly confronts the city's poor, their families soon became reconciled to their absence, as it serves to lighten somewhat the burden upon the breadwinners.'[31] But it's clear that some of the incarcerated were there through a simple lack of sleep and subsequent mental exhaustion.[32] Riis tells us of one tenement family who never made it as far as Ward's Island. They took poison together, he writes, simply because, living as they did, surrounded by constant disturbance, 'they were "tired"'.[33] These were people driven to a point of distress well beyond what, back in London, the Carlyles had termed being 'out of sorts'. For the people of the Lower East Side, noise really could become a matter of life and death.

In the late nineteenth and early twentieth centuries, the people of the tenements had to scramble for crumbs, and one of the scarcest resources was peace and quiet itself. They had little power over their own soundscape, and little chance to escape it. Since then, however, the twentieth century's commitment to social reform has changed this place out of all recognition; in the 1930s, especially, it made tenement life more bearable. But of course those old living conditions haven't entirely disappeared. They have simply moved. These days, the sweatshops that produce the world's cheap clothing are to be found in the Far East and the Indian subcontinent. And so it's there, too, that we find the noise, sleeplessness and nervous exhaustion which once so worried Victorian reformers such as Dan McKenzie.

Meanwhile, the bustle of a neighbourhood such as the East Village or the Lower East Side is still there to be heard, although in muted form. Today, Manhattan's bohemians, the young and the prosperous, come to savour its air of manageable edginess. These new residents enjoy the fact that there are always people near, that it's a place where you still see and hear each other, which draws, as Katherine Greider has written in her history of her home on the Lower East Side, on the 'informal, expressive social traditions' of its old working-class immigrants' – which, in short, offers just the right 'degree of intimacy'. Density, it seems, doesn't always lead to overcrowding. People have become adapted to it, largely, psychologists tell us, because, unlike their predecessors, they are not competing for scarce resources.[34] They can decide when to be private, when to be public, when to be stimulated, when to be relaxed. In short, as they have control over much of their lives, they have control over their soundscapes. As always, it makes all the difference in the world.

25

Capturing Sound

It's difficult to forget the sight of New York's Twin Towers collapsing on 11 September 2001. On the day itself, television was there to capture the sheer scale of the attack. Beamed live around the planet, then repeated endlessly in thousands of news bulletins, its footage meant that wherever we were in the world we, too, were witnesses to Manhattan's trauma. But it was through sound that we were taken to places well beyond the reach of the cameras – both physically and emotionally. Although television and photography provided some extraordinary visceral moments, it was through sound, more often than vision, that the most intimate human dimensions of the day's events really hit home.

One of the most haunting features of 9/11, for instance, was that many of those trapped in the towers or on the planes – people who often knew they were about to die – left messages on the telephone answering machines of their loved ones. There was the message from a man on one of the hijacked planes to his wife. 'I want you to be happy, I want you to carry on,' he said. 'See you when you get here.' Another was from a New York firefighter called Walter Hynes, to his wife, Veronica. 'I don't know if we'll make it out,' he said. 'I want to tell you that I love you and I love the kids.' Neither message is elaborate or poetic, but each is the last thing either woman heard of their husbands. And in the first year after his death Veronica Hynes played that recording of Walter back to herself hundreds of times. 'He was thinking about us in those final moments,' she says. 'That gives me great comfort.' In the years since, she has even made extra copies of the recording. Walter, it seems, will be heard by his children and, eventually, his grandchildren.[1]

Messages such as Walter's are now part of the 'Sonic Memorial Project', an online audio archive designed to help us remember 9/11 and the people who lost their lives that day. The Memorial's creators, Davia Nelson and Nikki Silva, claim the value of these bits of audio is that they provide 'immediate first-person accounts' of the event 'from almost every vantage point'.[2] But they are also a moving testimony to the sheer emotional hold such sounds still have over us. They are a reminder, too, of the extraordinary impact over the past 120 years of the apparently simple ability to preserve sound.

So far, I have only been able to discuss sound as something fugitive: there one moment, gone the next. People such as the Italian scientist Giovanni Batista della Porta might have fantasised back in the sixteenth century about the possibility of sealing sounds in lead pipes and releasing them later,[3] but this was just fantasy. In the real world, sound has remained stubbornly ephemeral. Until, that is, the last few decades of the nineteenth century. In 1860 the French typesetter Édouard-Léon

Scott de Martinville managed to 'print' the sound of a woman singing 'Au Clair de la Lune' by recording visual images of the sound; however, he couldn't work out a way of ever replaying it himself (in fact, it was not until 2008 that Scott de Martin-ville's recording was heard, when it was finally converted to a digital audio file by scientists in Berkeley, California). It was only after 1877, the year in which Thomas Edison invented a workable recording machine, that we were able not just to capture sound but also to replay it at will – 'I was never so taken aback in my life,' Edison said of the moment he first heard his own voice coming back to him.[4] And our reaction even today, when we hear the voice of someone dead, shows how sounds from the past still stop us in our tracks. When Veronica Hynes heard Walter's voice in 2001, it was as if for her he was alive again – a brief moment of contact. For us, too, it's an example of how recording has allowed sound to reach out and touch all sorts of people in new and surprising ways.

It probably seems rather morbid of me to begin with death. After all, the story of recorded sound is mostly a story of recorded music; in other words, a story of entertainment and pleasure. But in its earliest days a more serious future was being predicted for this technical discovery. Edison's preferred term for his invention – the phonograph – meant, literally, 'voice-writer'. It was a single machine that didn't just play recordings, it also made them. So it might become a handy dictation-machine for the office, a means of teaching elocu-tion or perhaps a tool of distance learning. More imaginative writers wondered whether it might even become a means of family surveillance. A short story published in 1888 suggested that if a phonograph were to be concealed under a sofa it could eavesdrop on scandalous behaviour: a father suspicious of his daughter and her boyfriend would then be able to con-front them with the evidence of what they had got up to.[5] But it was the promise of immortality that excited people most.

For Edison, this meant using the microphone to preserve people's languages, their speeches and especially 'the last words of a dying man'[6]: 'Centuries after you have crumbled to dust,' he wrote, '[the phonograph] will repeat again and again to a generation that will never know you, every idle thought, every fond fancy, every vain word that you choose to whisper against this thin iron diaphragm.'[7]

The phonograph, or, as most Britons soon learned to call it, the gramophone, was soon put to use capturing famous voices, for example that of the writer Robert Browning in 1889, or, the following year, of the nursing reformer Florence Nightingale. These recordings weren't exactly best-sellers, not least because mass reproduction wasn't possible yet. That came only with the arrival of shellac and the 78rpm disc a few years later. For now, in middle-class homes with a phonograph, people were more likely to record and listen back to their own family members. They would draw the parlour curtains, invite friends around, summon up the bodiless voices, then hear them float uncannily through the air. It was a kind of seance in which they got to choose their own ghosts.[8]

Arthur Conan Doyle once suggested that recorded sound offered 'a communion with the past',[9] but it offered more than that. Once captured, sounds became detached – freed, so to speak – from the people who had made them. So, as well as being made permanent, sound was now also made portable. Listening to a gramophone record was a chance to be reminded of another place as well as another time. When so many people were on the move – migrating from one country to another, moving from countryside to city, heading for the battlefield, posted to a distant corner of the empire – it offered a wonderful emotional link with home, as one young Englishman observed in 1910, while encamped in the Kenyan bush on his first trip to Africa and moved to write a poem in praise of his own precious gramophone:

Hark! It awakes and, like fettered bird,
Soars o'er the camp, till from its melody
Floats down the wordless longing for things loved
And English hearts are strangely nearer to me.[10]

It was music, not speech, that moved this young English-man. By the first decade of the twentieth century, technical improvements meant the quality of recording was just about good enough to capture musical performances as well as human voices. Indeed, it was soon perfectly natural to think of gramophones and music as belonging together.

But what kind of music would be recorded in the years ahead? The same old melodies and songs people had always played among themselves or listened to in concerts – or some-thing different? The question had to be asked, since there were signs that the very act of recording music was changing it. For a start, sounds could now be manipulated – repeated, distorted, speeded up, slowed down. So what if recording were to help unleash a completely new harmonic order, one that captured the spirit of the new age of cities and machines?

One group guaranteed to explore the outer limits of the possible was the Italian Futurist movement. Its leading light, Filippo Tommaso Marinetti, had declared in 1909 that the roar of the motor car was more beautiful than anything by Michelangelo. When it came to sound, however, the key member of the group was Luigi Russolo. In his Milan labora-tory Russolo constructed so-called 'noise-intoners' in order to capture the multitude of sounds in the world and create what he called the 'art of noises':

Noise … arriving confused and irregular from the irregular confusion of life, is never revealed to us entirely and always holds innumerable surprises. We are certain, then, that by selecting, coordinating, and controlling all the noises, we will enrich mankind with a new and unsuspected pleasure

of the senses. Although the characteristic of noise is that
of reminding us brutally of life, the Art of Noises should
not limit itself to ... imitative reproduction. It will achieve
its greatest emotional power in acoustical enjoyment itself,
which the inspiration of the artist will know how to draw
from the combining of noises.[11]

For Russolo, music had until now drawn inspiration only from
the narrowest spectrum of available sounds. Noise offered a
near-infinite range. Audiences, it turned out, were less convinced. *The Times* reviewed a London performance by Russolo
and his 'noisicians' in 1914:

After the pathetic cries of 'no more' which greeted them
from all the excited quarters of the auditorium, the audience
seemed to be of the opinion that Futurist music had better
be kept for the Future. At all events, they show an earnest
desire not to have it at present.[12]

If Russolo himself couldn't quite realise his vision, there
would be other avant-garde composers capable of revolution-
ising musical culture, by using noise in a more subtle fashion.
Stravinsky, Stockhausen, Antheil and, later, John Cage would
all owe a small debt to these pre-war Futurist experiments.

Meanwhile, other musicians found that the new technology of recording drew them back in time, rather than forward. In the years before the First World War, the Hungarian
composers Béla Bartók and Zoltán Kodály, for instance, went
foraging in the countryside to hunt down and capture on
disc the sounds of Eastern European folk music. Being able
to play these tunes back later and listen carefully over and
over again was, Bartók reckoned, like being able to put them
'under a microscope': features that had previously floated past
unnoticed – quarter tones, slides, subtle alterations in tempo
or tone colour – could now be pinned down.[13] Indeed, the

two composers were soon convinced that it was these aspects, so difficult to convey in music scores or catch at first hearing, which made music truly expressive. Increasingly, they infused their own classical compositions with the more fluid elements of folk. Indeed, their fusion of classical, folk and modernist styles was the kind of music which could have been composed only in the gramophone age.

The simple fact that sound was recordable and portable meant that from the 1920s the repertoire of music exploded in both volume and variety. Until then musical fashion had kept burying the past – except, of course, for the small canon of great works. But now the catalogue of music just kept growing; music from any period or any part of the world could be brought to life and folded into the present and future. It was as if music could go in any direction it wanted. Or, more accurately, *almost* any direction, for even in the 1920s the technology of recording still set firm limits on what worked. And, of course, there was money to be made from selling records. A disc made in Milan in 1904 by Enrico Caruso was the first to sell over a million copies, thanks in part to a large community of Italians in America who craved operatic excerpts from back home. Record companies quickly calculated the financial rewards in satisfying the tastes of that new social group, the record-buying general public.

This was the heyday of 'Tin Pan Alley', the name given to the collection of commercial music publishers located in New York City in the early years of the twentieth century, specifically on West 28th Street between Sixth and Broadway, although the phrase is really more of an idea than a place. These publishers had been selling sheet music for some time and doing very well out of it, but they began to dominate popular music in America when they responded energetically to the record industry's need for songs of a very particular kind. Everyone in the music industry knew that recording technology couldn't yet capture the full range of sound being made by an orchestra or dance band, or even by the normal singing voice: microphones picked up some instruments better than others; variations in shading and nuance were lost; and, anyway, the maximum playing time on the first discs was only about four minutes per side. All of this meant the need was for unfussy tunes, loud instruments and a strong but resolutely middle-of-the-road singing voice. There wasn't much room for subtlety.

Tin Pan Alley, then, wanted popular tales of love or betrayal belted out in three minutes and thirty seconds flat. Naturally, some musicians rose to the challenge better than others. Seasoned stage performers trembled when faced with having to sing into an impersonal recording horn, knowing their slightest imperfections would be immortalised. Those who adapted to the demands of the medium had it made. Billy Murray was one. He deliberately rounded out his vowels, and his tenor voice, it was said, had a certain 'ping' to it that cut nicely into the wax.[14]

Thanks to hundreds of writers and singers working in the spirit of Irving Berlin and Billy Murray, the commercial recording industry was itself a hit. By the 1930s, popular music on mass-produced discs brought familiar tunes – and a great deal of pleasure – to millions of homes in America and Europe. Tin Pan Alley's commercial model also proved to be highly exportable. In India, for example, one American company set up so-called 'training centres' to turn out popular recording artists. Local musicians were taught to set existing poems to music, thereby generating a steady supply of 2,000 or 3,000 new songs each year. It was exactly this music that would later become a strong influence on the Indian film musicals of the 1940s and beyond.[15]

The typical 1940s Indian film soundtrack, like the snappy music of Tin Pan Alley, was successful largely because it was reasonably predictable. And there was a price to be paid. By choosing to popularise singers whom its own talent scouts considered most suited to the medium, the commercial music industry, so Michael Chanan has argued, gradually 'robbed music of its sense of spontaneity'. Little by little, risky experimentation was excluded, and ever since then, proven harmonies and styles have become embedded as the standard musical building blocks of the commercial machine. Each new record has had to be different, yes. But not *too* different. In the age of recorded sound, Chanan concludes, 'everything is always new and always the same'.[16]

Yet if this is really the case, why is there so much variety in popular music, even today? Go into a good independent record store in London or New York or Tokyo or Paris – in fact in pretty much any city or large town – and you will be surrounded by CDs or vinyl boxed into endless categories and genres. And you will sense, too, the passion with which record-buyers, as members of musical tribes, sometimes cleave to one style or another. For them, clearly, popular music isn't all the same, nor is this variety just a western luxury.

The clearest example of this can be found in Ghana's capital city, Accra, which many music aficionados would regard as having one of the liveliest and most fertile musical cultures in the world as a result of its history and location. Its downtown record stores and market stalls are packed with music from all over the world, especially reggae, dance music and American hip-hop. There is plenty of African music, too: Afrobeat from Nigeria, bongo flava from Tanzania, kwaito from South Africa, soukous from Congo and, of course, Ghana's own specialities, highlife and hiplife. In Chapter 19 we learned how in Ghana the old slave-trade routes between West Africa and America were also routes followed by music. Back and forth across the Atlantic there has been what Halifu Osumare calls a 'circle of musical and dance influences from Africa to its diaspora and back again'.[17] In the 1920s, big-band jazz, gospel, brass band and sea songs – all of which owed something to African musical traditions – were imported from America, and blended to create the mellow sounds of highlife. In the 1990s, it was American hip-hop beats that were brought to Ghana, but instead of being adopted wholesale, they were blended with African rhythms and lyrics to create the new hybrid, hiplife. And, of course, over time – and as it's spread – hiplife, too, keeps mutating. Head to the north of Ghana, to the city of Tamale where Islam is stronger, and you will hear a more haunting, hypnotic version – a so-called 'melismatic' sound, reminding us, perhaps, of old Orthodox Christian as well as Arabic singing methods. Head to Britain or Germany, where Ghanaians have emigrated, and you will get another version: a harder sound, with funk and rock influences.

This process – one in which new hybrids are constantly being created and then evolving into new genres before they in turn are borrowed and adapted – has depended on one thing more than any other: the ability of musicians and music fans to listen closely to sounds that were made many hundreds of miles away and sometimes many years before. In other words, it has depended on their access to recordings – as it has for decades.

Just as Tin Pan Alley was busy turning out a stream of instantly catchy and inoffensive hits in New York's Midtown, other kinds of music were taking on a life of their own away from the spotlight. In the 1920s, many New Yorkers in Harlem would come to venues such as Big Joe's on 47th Street, or maybe travel to one of the 7th Avenue record stores just below Union Square, and meet friends, talk, listen to music and buy discs.[18] Many of these customers had recently migrated to the city from New Orleans and the South, bringing their music with them: blues, ragtime and jazz. Artists such as Mamie Smith, Ma Rainey and Bessie Smith filled these meeting places with a steady supply of what the music companies then called 'race records'. When it was released in 1920, for instance, Mamie Smith's 'Crazy Blues' sold at the phenomenal rate of 8,000 copies a week. Clearly, African Americans were by now a big and potentially lucrative market. There was, of course, an even bigger market to be tapped. The first jazz recording to sell over a million copies had, after all, been by the Original Dixieland 'Jass' Band – a group of five white musicians. So Tin Pan Alley found it easy enough to take what they thought of as the 'raw' music of Harlem and the South and use white imitators to transform it into a smoother, sweeter, less threatening sound for the mass market.

Over the following years, jazz music was ceaselessly on the move, regenerating itself through a mixture of slow adaptation and sudden bursts of creativity. And in most cases, it was recorded discs that provided the fuel for change.[19] People no longer needed to attend live performances in the clubs of Harlem or downtown to hear these new sounds: they could listen to them on disc in the record store or in the privacy of their own home, whether home was Harlem itself or the suburbs somewhere in Texas – indeed, whether home was African American or white. Neighbours and friends would play records to each other at each other's houses, at parties or picnics. They would turn on the radio and sometimes hear the latest releases. And, according to the American writer Harold Courlander, what they heard when they gathered and chatted and listened could be an inspiring jumble of sounds:

> … we may hear cowboy tunes that are reminiscent of Negro blues; blues that sound like the songs of the Golden West; hillbilly tunes … with jugs, jews-harps and washtubs; jazz-like treatments of old religious songs; Calypsoish skiffle bands in New Orleans and Mobile; and gospel songs with a suggestion of *Moon Over Indiana* in them.[20]

As Michael Chanan shows us, 'the music circulated far more

widely than simple statistics might suggest'.[21] And as it circulated it cross-fertilized, then split off again – because all this listening was bound to leave its mark. Many jazz musicians who grew up in the 1920s and 1930s have described how they developed their instrumental technique or singing style by imitating the records they heard.[22] Listening to recordings, just as closely as Béla Bartók and Zoltán Kodály had done in Hungary, meant learning about the music. It was a form of learning no longer tied down by the precision of a scoresheet or the conventions of racial identity: it was a looser, less predictable kind, which left more room for spontaneity and improvisation. After all, no attempt to copy what you heard on a record could achieve a perfect replica. Each imitation would be slightly different. Everything is always new, and *not* always the same.

The industry that grew up around recorded music in the twentieth century helped give the Age of the Machine its distinctive soundscape, just as much as the railways and factories of the Industrial Revolution had a hundred years earlier. In its own way it contributed to the background rhythm of city noise that was busy, human-made, repetitive – and which, increasingly, was thought of as ever-present. Often, the unceasing noise could be overwhelming, especially for those who for reasons of poverty had little control over any part of their own lives. Usually, though, there was too much human ingenuity around to allow one sound to dominate all others. Amid the din, people used their ears and found instead of one mechanical roar they were surrounded by an unexpected variety of sounds. They discovered new dimensions to music, the sublime in nature – even, thanks to the stethoscope and the microphone, the marvellous hidden rhythms of their own bodies. Sometimes, they even created completely new sounds with which to entertain themselves. Organic, natural, human-scale noise had once seemed under threat of extinction in the face of an all-consuming industrial noise. In the end it survived. How it would fare in the face of electricity and amplification, in the face of a globalised mass media and in the face of the most mechanised warfare the world had witnessed is the subject of the final part of this book.

VI
The Amplified Age

26

Shell Shock

On 7 June 1917 the British army detonated a series of enormous mines they had placed in tunnels secretly dug right underneath the German army lines on Messines Ridge near Ypres in Belgium. It was a common enough tactic to set off mines, but on this occasion the sheer magnitude of the blasts was extraordinary. When they started in the early hours, it was said they could be heard clearly by the Prime Minister David Lloyd George in Downing Street, some 140 miles away. According to some, they could even be heard in Dublin.[1] For people living near the Kent coast these devastating explosions would have been still louder, but for these civilians the sounds of battle were already familiar: in this part of England the distant rumble of artillery bombardment in northern France had been heard almost daily for more than two years now. And what they had heard throughout this time was merely a hint of the noise that would have been assaulting the hundreds of thousands of men at the Front itself.

Among these men was the writer Robert Graves, serving with the Royal Welch Fusiliers. In his memoir of the war, *Goodbye to All That*, he describes how 'the noise of the guns grew louder and louder' as he had approached the Front for the first time. Here, Graves found himself in a war fought on a massive, industrialised scale. The Industrial Revolution had once threatened to turn humans into mere cogs in the machine, and those caught up in the conflict in Europe seemed to confirm that this nightmare had now come true. The sounds of the battleground seemed to symbolise for soldiers such as Graves a puny helplessness in the face of something monstrous and out of control:

> A colossal bombardment by the French at Souchez, a few
> miles away – continuous roar of artillery, coloured flares,
> shells bursting all along the ridge by Notre Dame de Lorette.
> I couldn't sleep. The noise went on all night. Instead of
> dying away it grew and grew, till the whole air rocked and
> shook … I lay in my feather-bed and sweated. This morning
> they tell me there was a big thunderstorm in the middle of
> the bombardment. But, as Walker says: 'Where the gunder
> ended and the thunder began was hard to say.'[2]

If, like Graves, you were at the Front, you didn't so much hear the noise as feel it. It went right through you, shaking you to the bones. But among the messy complex of trenches and dugouts that stretched for hundreds of miles along the countryside of northern France, what really made sound a defining experience of battle was that so much of the war could never be seen. Paul Fussell described the trenches as a 'troglodyte world', where sunrise and sunset took on extra significance simply because the sky was visible like nothing else. It was obviously risky to put your head above the parapet. Even if you did, the chances were that, with all the barbed wire, the mounds of earth, the smoke, the explosions and the gunfire, you would still be in a largely sightless world.[3]

Crouched down low in the mud, however, you would still have been able to *hear* a great deal. It was your ears that were your connection with what was happening above ground. In *All Quiet on the Western Front*, Erich Maria Remarque wrote how 'Every man is aware of the heavy shells tearing down the parapet': the furious blasts were like 'a blow from the paw of a raging beast of prey' somewhere just out of view.[4] Although his is a fictionalised account, it's based on first-hand experience, and Remarque's work, like Graves's – indeed like countless other reports from those who had witnessed the slaughter – is full of vivid descriptions, not so much of the sight of war as its sounds: shells howling and whizzing and hissing as they fly past, the crashing noise of metal casings falling to the ground, the rattle of machine guns, the crack and ping of rifle bullets, artillery booming through the night like a church organ or bellowing in the distance 'like a rutting stag'.[5] It was no coincidence that the German war poet August Stramm chose to construct his poem 'No Man's Land' around a chain of sounds:

Grausen
Ich und Ich und Ich and Ich
Grausen Brausen Rauschen Grausen
Träumen Splittern Branden Blenden
Sterneblenden Brausen Grausen
Rauschen
Grausen
Ich

[Dread
Me and me and me and me

Dreading roaring crashing dreading
Dreaming splintering burning dazzling
Dazzling star-shells roaring dreading
Crashing
Dread
Me]⁶

Trench warfare brought an onslaught of noise. Yet it seems that not all noise was the same. And variety gave it meaning. Soldiers learned how to discern different and distinct sounds amid the din, and make sense of what they heard to work out what might be happening around them. This took time. In 1915, for instance, Graves had found himself in the midst of a big push, trying desperately to figure out the progress other units were making further down the line. All he could hear was 'a distant cheer, confused crackle of rifle fire, yells, heavy shelling on our front-line, more shouts and yells, and a continuous rattle of machine-guns': a jumble of sounds almost impossible for a new recruit like him to interpret.⁷ But in 1915 Graves was raw material. As he became a more seasoned fighter, like his comrades and his enemies he discovered how to tune his ears, and how he needed to react differently to different noises.

Take gunfire. Graves learned that the sound of a rifle bullet was particularly alarming, simply because the bullet travelled too fast to give any advance warning. Soldiers like him learned not to waste effort ducking for cover: by the time they heard it they knew it must already have missed them. Yet even if a rifle was being fired blindly, it somehow seemed 'purposely aimed', and its sound therefore gave off a horrible feeling of real personal danger.⁸ The rattling of machine guns was different. Ludwig Scholtz, a medical volunteer on the Eastern Front, described it as inspiring not so much a feeling of personal danger as a pervasive 'sense of foreboding', the rattling announcing that a more indiscriminate form of attack was under way.[9]

The major soundtrack of the war, however, was the booming noise of artillery. It was, one British officer said, 'the master of the battlefield', its thunder 'like a gigantic force that never stopped, day or night'.[10] However, although it was devastating in scale, at least it sounded as if the enemy's gunners were firing not at people but at 'map references' or target areas.[11] Artillery also gave men some chance of avoiding injury. As Graves writes:

> One day, walking along the trench at Cambrin, I suddenly dropped flat on my face; two seconds later a whizz-bang struck the back of the trench exactly where my head had been ... The shell was fired from a battery near Les Brigues Farm, only a thousand yards away, so that I must have reacted simultaneously with the explosion of the gun. How did I know that the shell would be coming my way?[12]

The answer, of course, was that the sound of the gun being fired had travelled faster than the shell itself, giving advance warning of its arrival. By now Graves knew how to react instinctively, the mark of an experienced soldier, which he believed anyone would become after only a few weeks. By that point, he claimed, 'We can sort out all the different explosions and disregard whichever don't concern us'.[13] Indeed, bombardment was sometimes so continuous that it became a mere backdrop – a noise that could generally be ignored, even, for some, slept through. This allowed soldiers to focus semiconsciously on those sounds truly worthy of a response – such as the shout to 'Stand to' – and then snap awake just in time.[14]

However, no soldier could afford not to be alert to a subtle shift in the background sound of artillery, since it would have provided valuable information. Specially trained military officers on both sides spent a great deal of time 'sound-ranging', using detailed mathematical calculations to work out the precise location of enemy gun positions based on the time lag between firing and shell-burst observed from different perspectives. If they detected a changing pattern, it meant the enemy might be on the move or preparing an offensive. Ordinary soldiers developed a rough-and-ready kind of sound-ranging of their own to help avoid the worst of enemy machine-gun fire or locate the weakest links in enemy defences as they moved about the field of battle.[15]

However, there was one noise that traumatised a soldier more than any other as he sat hunched in his trench or was lying low in a shell-hole: the gut-wrenching sound of a comrade dying. Once, when Robert Graves was trying to catch some sleep, he heard a man lying on the floor of the trench, 'making a snoring noise mixed with animal groans'. He had lost the top part of his head, a fatal injury, but he still took three hours to die.[16] Nor would he have been the only one suffering this very public and horribly drawn-out end. The German poet Anton Schnack described the scene in the aftermath of one battle:

> I am lying here amid murder and assaults, in a blue sea of
> rockets, in the wind's sighing, beneath turbulent night skies,
> in green waters full of snails and red worms, awaiting death,
> putrid and swollen, amid the dying screams of horses, amid
> the dying screams of men, I heard them, calling out of the
> dark, hanging in the wire; thus do birds sing who are ready
> to die, lonely, pining away, in the spring of their lives.[17]

It would take a brave man to go over the top, pull someone from the wire and then bring him back to safety; a dying soldier might be close, yet still out of reach. The brutal reality was that it was one's duty, if injured, to go as quietly as possible. However, as this account by Graves of one soldier's struggle to die in this selfless fashion shows, it wasn't easy:

Samson lay groaning about twenty yards beyond the front trench. Several attempts were made to rescue him. He had been very badly hit. Three men got killed in these attempts; two officers and two men, wounded. In the end his own orderly managed to crawl out to him. Samson waved him back, saying he was too riddled through and not worth rescuing; he sent his apologies to the company for making such a noise.[18]

At dusk, Graves went into no-man's-land and found Samson's body. It had been hit in seventeen places: 'he had forced his knuckles into his mouth to stop himself crying out and attracting any more men to their death.'[19]

Alas, the sound of death would not have ended, even then. In *All Quiet on the Western Front*, Remarque describes the sickening noise of corpses – the hissing and belching as they released their gases. It would have been hard to forget. The writer William Faulkner certainly didn't. What he called the 'little trickling bursts of secret and murmurous bubbling' emitted by the shattered bodies that lay around him on the battlefield stayed with him for years.[20]

The sound of death, the screams and groans, the roar and rattle and howl and crack of gunfire, the anxiety and uncertainty each time a shell was heard heading your way: none of this could be experienced for any length of time without having some impact on one's morale or nerves. As one British medical officer suggested in his diary entry from April 1916, this sensory assault was 'entirely outside' the normal experience of any man. 'I do not know,' he added, 'how the Infantry stand its continuity … and still stay sane.'[21]

The answer was that many of them didn't. Take the case of Private George Gomm of the 22nd Battalion, Royal Fusiliers. In April 1916 Private Gomm had been found wandering around near the trenches in a deeply traumatised state, having left his company. He claimed he had 'had a shock'. But was this a sudden physical shock, from which he might recover? Or might it be a deeper psychological shock, the effect perhaps of a slow, steady wearing down of his mental resources? Might it even be a case of malingering? 'Shell shock' was a matter of heated debate, even during the war itself, and although no group of army medical officers would ever fully agree, in most diagnoses of cases like this, exposure to sound kept cropping up as a possible cause.

Edwin Ash, a neurologist working in London, had warned at the outset of war that the noise of battle would be a 'powerful factor in shattering the nerves'. The vigorous battering of the eardrums and auditory nerves caused by a shell exploding nearby, he suggested, would inevitably trigger 'a corresponding vibration in the brain'.[22] Charles Myers, an army medic plucked from a research laboratory in Cambridge and now working in military hospitals across northern France and Flanders, agreed that for some soldiers there was indeed very real physical damage of this nature, leading to deafness, loss of memory, even shaking and disorientation, but that this was no more than an extreme form of concussion, something from which they would probably recover.[23] What concerned Myers much more were the hundreds of other cases he had witnessed where the same symptoms had appeared yet the patients had apparently been nowhere near exploding shells. In these cases, he believed, the shell shock was more likely to be a result of the cumulative effects of battle. These were men, perhaps, who had simply been there too long, who had reached the limit of what they could bear of the horror and the fear:

> ... in men already worn out or having previously suffered from the disorder, the final cause of the breakdown may be so slight, and its onset so gradual, that its origin hardly

deserves the name of 'shock'. 'Shell shock', therefore, is a singularly ill-chosen term.[24]

Myers doesn't talk much about noise in his diagnosis, but if we dig deeper into his case studies it turns out that noise still mattered more than anything else. Although he was sure that, among the 2,000 or so cases of shell shock he had examined by the middle of 1916, some were the result of men having had psychological problems before the war, many other soldiers had succumbed simply because they were worn out. They were worn out because they were not getting enough sleep, and they were usually not getting enough sleep as a result of constant noise and interruptions. Robert Graves estimated that at one point he was getting no more than eight hours' sleep in every ten nights. This, he suspected, was typical. Yet he knew that sticking it out like this came at a high price:

Having now been in the trenches for five months, I had passed my prime. For the first three weeks, an officer was of little use in the front line; he did not know his way about, had not learned the rules of health and safety, or grown accustomed to recognizing degrees of danger. Between three weeks and four weeks he was at his best, unless he happened to have any particularly bad shock or sequence of shocks. Then his usefulness gradually declined as neurasthenia developed. At six months he was still more or less all right; but by nine or ten months, unless he had been given a few weeks' rest on a technical course, or in hospital, he usually became a drag on the other company officers. After a year or fifteen months he was often worse than useless.[25]

Graves, a member of the officer class, naturally focused on the travails of his fellow officers. But some two decades after the end of the war in Britain, according to the historian of psychology Peter Barham, 'there were around 35,000 war disability pensions still in payment for ex-servicemen with mental disabilities', and 'the vast majority of these, quite disproportionately so, were other rankers, ordinary soldiers rather than officers'.[26] These were men such as Private Martin O'Brian, a former baker from Stafford, who had lost his reason through shellfire and told doctors that he had 'not been sleeping well since due to hearing the noises of shells and guns'.[27] And Private Charles Butterworth, who, after being injured in a blast at Armentières in December 1915 became hard of hearing and increasingly mute, but who also had recurring nightmares of being in the trenches and hearing the shells with a 'noise as of gas' in his ears. He later woke up in hospital one night, very agitated, shouting that everyone 'must take cover from the shells which were coming over'.[28] These were men who had not just been tormented by sounds in the battlefield, but were *still* being tormented years later by imagined sounds in their heads. Just as much as Graves's officers, they were men broken by noise.

Most doctors agreed that if one likely cause of shell shock was too much exposure to noise, the obvious treatment should be to offer its opposite: plenty of peace and quiet. Yet the treatment these men received when they were sent back home depended on their social status, which meant they got radically different amounts of relief. Officers, for instance, might be sent to a place such as the Special Hospital for Officers, which opened at Palace Green in Kensington in January 1915. The neurologist Edwin Ash gives a hint of the underlying thinking here:

> ... our officers, whose brains are on the whole more highly trained and more delicately organized than the physically stronger soldiers of the line, should be cared for by proper rest and mental support when suffering from nervous shock ... Unless proper precautions of this kind be taken many

valuable young brains will be harmed and their services lost
to the nation.[29]

Naturally, the ideal environment for these valuable young
brains would be deep countryside, since, as the historian
Fiona Reid shows, 'the antidote to a brutal industrial war
appeared to lie in the creation of a rural idyll'.[30] But Kensington at least had its own garden, and a detached building
for a maximum of thirty-three patients, each with his own
room. Here, serenity reigned.[31] For as Edwin Ash explained,
in treating patients, 'no drug, no cure, no operation will
help so much as a few hours' natural sleep' in a 'quiet, dark
room'.[32] If any sound dared intrude, it would probably only
be music coming from the treatment room downstairs, since
it was in a hospital such as this that music therapy experienced
a wartime revival: it was recognised once more as a powerful
means of helping patients regain a flowing sense of physical
movement. In this, and every other respect, Kensington was
a model of best practice.

Unsurprisingly, the families of rank-and-file soldiers suffering from shell shock wanted the same for their men. But while
an officer was likely to be seen as a 'neurasthenic' who would
clearly benefit from a few weeks' rest, the ordinary soldier
was more likely to be seen as intrinsically weak – a 'hysteric'
beyond treatment. So Private George Gomm, whom we met
earlier wandering around northern France in a deeply traumatised state – someone whom Charles Myers himself regarded
as eminently treatable – ended up on his return to Britain
not in Kensington but in a large military hospital in South-
ampton. There he was ticked off for having 'no manners',
declared 'foolish and stupid' and finally dispatched to a lunatic
asylum, where he would be forgotten.[33] Other rank-and-file
soldiers, even if they didn't end up incarcerated like Gomm,
usually found themselves in hospitals with a thousand or more

patients. These were busy, overcrowded and noisy: there was next to no chance of a rest cure. Whatever the good intentions of medical staff, the men's night terrors were likely to continue untreated.[34]

When the First World War was over and countless memorials had been built, it was perhaps fitting that a moment's silence was deemed the best way to initiate acts of remembrance. It was an obvious antithesis to the monstrous, violent din of conflict: a moment, almost liturgical in nature, when people were allowed the peace and quiet they needed to reflect or grieve in their own way.[35] Inside the heads of many soldiers, however, the noises of war reverberated long after Armistice. Edwin Ash continued to treat patients at his London hospital who still complained of 'buzzings and singings in the head'.[36] Meanwhile, Robert Graves, who saw himself as reasonably undamaged by the war, still complained that 'Shells used to come bursting on my bed at midnight'. It wasn't until 1928 that his flashbacks, which kept frustrating any attempt to catch up on what he calculated to be 'about four years' of lost sleep, finally left him.[37]

The war also left behind a much broader sensitivity to noise and its psychological effects. It was as if everyone's ears had been opened up to the profusion of sounds in civilian life. According to Yaron Jean, for instance, people in Weimar Germany frequently 'compared noises in big cities like Berlin with the ominous roar of yesterday's battlefields'.[38] In Britain, too, there was talk of a new urban soundscape of man-made sounds almost as threatening as those heard at the Front. Both, equally, seemed to represent sound of such scale and amplification that it appeared to be something moving beyond human control. Noise was now symbolic, not just of machinery or war, but of a whole runaway modern world in which people had less and less control over their own destiny. For a neurologist such as Edwin Ash, mental disorder was no longer a disease of soldiers, it was a 'national ailment'.[39] And all the amplified sounds of city life – loudspeakers, nightclubs, record players, radio sets – would become the next battleground in the twentieth century's war of nerves.

27

Radio Everywhere!

On the eve of the First World War, the Earth's atmosphere was crackling with the noise of communication – not just the dots and dashes of Morse code messages, but brief bursts of talking and music. Our own sounds were starting to be carried for the first time by radio. And for the tiny handful of people who had the equipment to pluck these faint murmurings out of the ether it was like witnessing a new kind of magic, as an excited young reporter explained in the *London Standard* on 28 December 1912:

> Mysterious music and voices in the air have puzzled
> hundreds of persons in England this Christmastide. From
> the silence of the night there have been wafted scraps
> of melody and tantalising fragments of conversation …
> Instead of regular beats upon the eardrum there have come

harmonic chords and short, unmistakable records of the
human voice.[1]

In these early days, listening to radio was an even uncan-
nier experience than listening to a gramophone. It had been
strange enough to hear someone's voice coming from a disc;
for it to come out of thin air seemed, well, *incredible*. Not
just out of the air, either, but from somewhere tens, hundreds, perhaps thousands of miles away. All these disembodied
voices were whizzing about, invisibly, in all directions – over
the horizon and across oceans, through solid buildings and
into living rooms – all at the speed of light.[2]

By the 1930s the novelty, and the magic too, had ebbed
away. Radio had become a normal, taken-for-granted, enjoyable part of everyday life. People gathered around radio sets
in cafés and bars, to drink to the sound of popular dance tunes
or catch the commentary for a vital boxing match or football
game. Millions sat by their fireside and listened to the news,
shopping tips, cookery advice, a thriller or their favourite comedy show. Yet although radio had woven itself so intimately
into the fabric of daily life that it was now barely noticed as
anything special – perhaps precisely because of this – its influence was stronger than ever.

Throughout this book I have discussed how sound has been
used as a kind of 'touching' at a distance – how, for example,
ringing bells have extended the reach of a parish church to
the furthest corners of a village, or how gunshots and canons
have helped put the stamp of royal authority over a remote
colony. In a very real sense, being within earshot of a sound
was what made you a citizen or subject. With radio, the distances involved were dramatically transformed. A single sound
made in front of a microphone placed on a table in a studio
in London, Berlin, Moscow or New York could be hurled
from the top of a giant transmitter and, in one instant, reach

millions. Indeed, each voice, each piece of music, each message – potentially, at least – had a national audience. Henceforth, a whole nation was in earshot. Radio, then, was intimate and enjoyable; just a box in the corner. But it was also the most powerful megaphone yet invented: manna from heaven for propagandists who wanted to mould an entire country to their way of thinking. And in the 1930s it was by no means clear which of these characteristics would be more influential in shaping the medium's future. Indeed, at various points in the decade, it was as if radio were handing dictators a loaded gun.

It was perhaps natural that the Nazis adored radio. When they seized power in 1933, they received 37 per cent of the vote in March and nearly 44 per cent in November, but wanted 100 per cent of the people behind them. And if the German nation was to be transformed into 'one people', the obvious means of establishing this unity of spirit was surely through broadcasting.[3] Cinema and newspapers were useful for spreading ideology, but the Nazi's propaganda chief, Joseph Goebbels, always pinned his greatest hopes on radio because of its unique ability to send a single, clear, unambiguous message from the centre to the furthest peripheries of the nation.[4]

Within days of being appointed, Goebbels summoned the various regional heads of German broadcasting to Berlin, and, pounding his lectern hard, set out his vision. Radio, he told them, was 'by nature authoritarian'; it 'automatically offers itself to the Total State':

> We make no bones about it: the radio belongs to us, to no one else! And we will place the radio at the service of our idea, and no other idea shall be expressed through it.[5]

To make this boast a reality, thousands of pillars were put up on street corners and squares, each one festooned with loudspeakers. Millions of affordable radio sets – the so-called

Volksempfänger, or 'People's Receivers' – were also put on sale, many of them specially adapted to make foreign broadcasts difficult to receive. By the outbreak of war in 1939, seven out of every ten homes in Germany had a radio set, the highest proportion of any country in the world. By this time, too, listening to radio was something of a national duty. On special occasions, such as a speech by Hitler, radio wardens would arrange for a siren to be sounded, and people would be expected to stop whatever they were doing and gather round a loudspeaker or radio set in a café, public square, school, office or on a factory floor.[6]

The Nazis' big mistake, however – at least to begin with – was forgetting that the German people had already got used to radio as a source of homely relaxation and pleasure. What they now got in place of their favourite shows was an unrelenting stream of political programmes or speeches by party officials, few of them as charismatic as the Führer. The low point came on the 'Day of National Labour' in May 1933 when, having been cajoled into listening all day, millions had to endure some twelve hours of non-stop speeches and political commentary, with no musical interlude whatsoever. Their evening treat was a two-hour speech from Hitler himself. Not only was this boring for listeners, it was hectoring, as if they were being harangued loudly in the privacy of their own home.[7]

The Nazis' gut instincts about radio's potential had been right, though: here indeed was the best means of getting a single viewpoint distributed across a whole country in a flash. And if their embrace of the new medium had been more like a vicious stranglehold, they were certainly not alone in their enthusiasm for radio's long-distance reach. In Soviet Russia, for example, Leon Trotsky had already advocated radio, not as 'a plaything for the privileged circles of the townspeople', but as a vital tool for converting a vast country of peasants, many of whom couldn't read newspapers, into a modern communist state – perhaps even for spreading the revolutionary message beyond the state's own borders:

> In order to introduce it we must first and foremost be able to talk to the most remote parts of the country ... The conquest of the village by radio is a task for the next few years, very closely connected with the task of eliminating illiteracy and electrifying the country, and to some extent a precondition for the fulfilment of these tasks ... Place the map for a new war on the table! ... It is necessary that on that day when the workers of Europe take possession of the radio stations, when the proletariat of France take over the Eiffel Tower and announce from its summit in all the languages of Europe that they are the masters of France, that on that day and hour not only the workers of our cities and industries, but also the peasants of our remotest villages, may be able to reply to the call of the European workers: 'Do you hear us?' – 'We hear you, brothers, and we will help you!'[8]

We might imagine that America, with its robust spirit of individualism and distrust of central government, had a very different approach. And indeed so the rhetoric went. In 1933, one industry spokesman claimed, not untypically, that the 'American air' offered 'all feasible liberty of utterance' to its people.[9] It certainly offered more entertainment than was allowed to the Germans in the early 1930s. A comedy series like *Amos 'n' Andy* was immensely popular – heard by something like 40 million Americans nightly at its height between 1929 and 1931. But of course the reason so many people heard it was that most of the hundreds of individual radio stations in cities and towns across the States were by now affiliated to big networks such as NBC or CBS, and so carried more and more of the same syndicated network programming. Although there is no doubt that this was led by audience demand, it was still a case of radio extending a cultural experience to an entire nation; you might say even helping to *create* a stronger feeling of nationhood. The same dialogue, the same storylines, the same gags were transmitted into homes stretching from the Atlantic to the Pacific coast, and millions of Americans were laughing out loud at exactly the same time as a result.[10]

The opportunity this presented to speak directly to the American people about altogether more serious matters appealed strongly to President Roosevelt. In 1933 he needed to get round the opposition of most newspapers and enlist public support for his New Deal plan to lift the country out of depression; a few years later, he needed to persuade fellow Americans towards a less isolationist stance over the war in Europe. In both cases, radio was the most powerful tool to hand. His most celebrated appearances were his 'Fireside Chats' – twenty-nine of them in all during his presidency – but he also made fifty other speeches on network radio in his first year of office alone. In every case, his aim was to use the power of human speech – amplified, so to speak, by radio – in order to change minds and shape public mood on the national scale.[11]

In Britain, under its founding father John Reith, the BBC was quite brazen about wanting to shape the public mind – to educate people, make them more cultured and enquiring; in short, to make them fit for a democracy and bound firmly into national life. Reith and his fellow BBC pioneers reckoned they knew what was 'the best' in the world of arts and ideas, and wanted to ensure the best no longer belonged only to a privileged few. In his 1924 book *Broadcast over Britain*, Reith wrote:

> As we conceive it, our responsibility is to carry into the greatest possible number of homes everything that is best in every department of human knowledge, endeavour and achievement, and to avoid the things which are, or may be,

> hurtful ... the organisation is being conducted on the lines
> of a public service, the maximum benefit to the maximum
> number being kept in view.[12]

For example, when it came to the pronunciation of English on air, Reith was clear that people's lives and employment prospects would be improved if an example could be set. So although there were countless regional dialects and accents to be heard in the streets, pubs and family homes of Britain, the 'standard' of spoken English to be heard on air should be what the BBC called 'southern, educated English'. 'So long as the announcer is talking good English, and without affectation,' Reith announced, 'I think it is much desired that he should be copied.'[13] The BBC saw radio as a way to bind a nation together through sound. If that meant not worrying too much about what audiences wanted, erasing a few regional differences and fiercely protecting the Corporation's monopoly, so be it. Democratic ends would be pursued through undemocratic means.

I'm not suggesting that Reith and FDR were totalitarians like Hitler, but what they certainly all shared when it came to radio was a keen awareness of what we might call its 'equalising force': its extraordinary ability in these years before television to distribute ideas across all classes and sexes and age groups, among cities and in the countryside, near and far. More than that, they realised that since everyone was listening to the same things at exactly the same time, radio stimulated a sense of *shared* experience: listeners, though they listened privately in their own families, were always aware that other people around the country were listening too. This didn't just allow them to imagine the nation, it helped them feel as if they were now playing a full part in its political and cultural life.

The risks in all this, though, were obvious. The dream of the Nazi radio programmers was to create a kind of

Volksgemeinschaft – a social unity that amounted, in effect, to a single way of thinking among the German people. British intellectuals worried that radio would create a uniform, 'middlebrow' British mentality. Americans, especially those on the Left, worried that, given commercial radio's desire to win the biggest audiences possible – and therefore reap the biggest advertising revenues possible – each competitor would end up creating the same range of bland, middle-of-the-road programmes. As one study of American radio's influence put it in 1935:

> When a million or more people hear the same subject matter, the same argument and appeals, the same music and humor, when their attention is held in the same way and at the same time to the same stimuli, it is psychologically inevitable that they should acquire in some degree common interests, common tastes, and common attitudes. In short, it seems to be the nature of radio to encourage people to think and feel alike.[14]

Radio was indicted as the great sonic homogeniser. But was this fair? Were audiences really starting to succumb to a kind of mass delusion or hypnosis through radio?

A small group of psychologists based at Harvard University in the 1930s were determined to find out. The two crucial figures were Hadley Cantril, a young researcher who had recently completed his PhD on public opinion, and one of his mentors, Gordon Allport, who was interested in how ordinary people might start thinking in authoritarian ways. Over several years, Cantril and Allport regularly gathered a group of human guinea pigs into the Harvard Psychological Laboratory, sat them down and exposed them to a series of fascinating radio experiments.

One involved them listening to various radio presenters addressing them through loudspeakers and then being asked what impression they had got about each presenter's personality. This being radio, the impression would have been based purely on the presenter's voice. And what the two psychologists discovered was that listeners formed a very strong mental impression indeed as to what these presenters were really like – their mood, their politics, even their appearance. The accuracy of these judgements was a bit hit and miss, but the point was that these disembodied voices were invested with personality.[15]

A second set of experiments divided the human guinea pigs into two groups: one that would read a printed text, another that would hear it read for them over loudspeakers. The findings this time were even more startling. Those who had read the piece were more critical and questioning about the material; those who had heard it over the loudspeakers were more inclined to believe everything that had been broadcast. Cantril and Allport were taken aback. Even if it is heard coming out of a box on the wall, they concluded, 'the human voice is more interesting, more persuasive, more friendly, and more compelling than is the written word'.[16]

This, then, was radio's great triumph: even though the words it broadcast were disembodied, came from hundreds of miles away and were addressed to an anonymous mass audience, they somehow still felt to the listener at home as if they were addressed personally to him or her – and by a real person, someone who was *there* in the same room.

For radio to have this effect, however, it couldn't act like a megaphone, shouting at people as if they really were bit-part players in a 'mass' audience – which is why John Reith never used the word 'mass', and rarely even used the word 'audience'; he talked of 'the listener', a singular person. And it's why FDR, in his 'Fireside Chats', always managed to sound as if he were talking one-to-one. He tended to speak slowly during these broadcasts, using direct phrases addressed to 'you', and hesitating sometimes; in other words, he spoke as if he were having a conversation.[17] The communications scholar Greg Goodale has suggested that Roosevelt's cadences ensured that every word 'fell upon listeners like a sledgehammer',[18] but that, perhaps, underplays the subtlety of his technique, which was really to sound as if he didn't have a technique at all. *The New York Times* was probably closer to the mark in its review of 18 June 1933:

> The human voice, when the man is not making conscious use of it by way of impersonation, does, in spite of himself, reflect his mood, temper and personality. It expresses the character of the man. President Roosevelt's voice reveals sincerity, good-will and kindness, determination, conviction, strength, courage and abounding happiness.

Most listeners certainly thought so. Thousands upon thousands sent him letters, talking back to their president as if he were someone they knew: 'If I am addressing you too informally,' one admirer wrote:

> it is only because you have brought yourself so close to us, the people. You may believe me that as I listened to you last night your every cough made me wince, and prompted me to admonish you, as I would one of the family, to take good care of yourself, for the country's sake, as well as your own.[19]

As for the Nazis, they had a tendency to shout. But Goebbels, to give him his due, knew that, no matter how good they sounded in the emotional cauldron of a stadium, endless hectoring speeches and martial music were utterly hopeless for the radio set in the family living room. And he told the broadcasters under his control as much: 'Whatever you do, do not broadcast tedium, do not present the desired attitude on a silver platter, do not think that one can best serve the national government by playing thunderous military marches every evening.'[20] As time went on, Nazi radio became an altogether smoother operation, complete with variety shows, dance music and chat.

This entertainment was never laid on for its own sake: under Hitler, it was always a means to an end – namely, integrating citizens into the German 'national community'.[21] Nevertheless, the softer style mattered profoundly. The psychologists at Harvard, the BBC, Roosevelt, even, in the end, the Nazis, had all discovered that radio's power was subtle and had to be handled carefully. Yes, it could speak to an entire nation at once, but not as a megaphone. If it was to seep into listeners' minds, it had to ingratiate itself with them. It had to swap the sound of the demagogue for the sound of the PR person or salesman. At the very least it had to sound like a companion. That's why, when television came along, and radio shifted from the living room into the kitchen and the bedroom, its power didn't diminish, it grew: there in the background, it became an important, ever-present soundtrack to our lives. Its voices and tunes and stories have continued to seep into our consciousness without us even noticing.

This might sound as if the dictators and salesmen won in the end. But they have always had to meet listeners largely on the listeners' own terms. The broadcasters' voices reach us in *our* territory, not in theirs. And if they try to seduce us, we are always free to remain blithely unmoved, just like one family in South Wales whose listening habits were observed by the BBC in 1938. As a Toscanini concert drifts out of the radio set in their living room, the woman of the house is seen sewing, waiting patiently for something she would much prefer to hear. Meanwhile, father and son are having a row, and other members of the household are drifting in and out, asking what the row is about. Neighbours are calling by, discussing the latest gossip without lowering their voices. The one solitary listener 'groans inwardly'. He doesn't, however, turn the radio off.[22] And I suspect that if he ever did, everyone else in the room would very quickly shout at him to turn it back on again. Even if they are not listening with their full attention, they are still listening. Even in the background, radio's burble is still more than a meaningless noise. And if it weren't there, they would miss it like hell.

28

Music While You Shop, Music While You Work

I n 1920 the French composer Erik Satie was having lunch with the painter Fernand Léger at a restaurant in Paris. The meal wasn't going well. The two men wanted to talk amiably to each other as they ate. But the restaurant had a resident orchestra, and so loud and enthusiastically was it playing that all their efforts at conversation were frustrated. They abandoned their meal and left. The occasion hadn't been a complete waste of time, however. Satie started thinking about what had happened, and told Léger of a possible solution for the future:

> You know, there's a need to create furniture music ... music that would be part of the surrounding noises and that would take them into account. See it as melodious, as masking the clatter of knives and forks without drowning it completely,

without imposing itself. It would fill up the awkward silences that occasionally descend on guests. It would spare them the usual banalities ... it would neutralize the street noises that indiscreetly force themselves into the picture.[1]

Satie decided to act. That same year he created a piece of furniture music of his own, to be played in the foyer during the interval of a play. It contained fragments of popular melodies, but remixed and repeated over and over again, a bit like Gregorian chant or the pattern on a piece of wallpaper, creating an aimless, almost hypnotic sound. As soon as they heard it, the theatre audience stopped talking and stood in reverential silence. Satie was furious. He leaped into the crowd, pleading with people to start talking, to make a noise of some sort or at least to spend their time looking at the pictures being exhibited on the wall. It didn't really matter what they did, as long as they took no notice of his music.[2] They should hear it, yes, but not *listen*.

In 1920 this was wishful thinking from Satie. His deliberate attempt to compose what was in effect sonic wallpaper upended almost every assumption about the relationship between music and listening. Well-brought-up Europeans had been told endlessly since the eighteenth century that music was something to be listened to with devout attention. Music was supposed to have character, to be full of meaning and emotion, not some noise to be ignored or even a sound to be heard half-heartedly. Indeed, given the amount of real noise thrown up by the world at large – from the industry and machinery and traffic of frenetic city living – it seemed more important than ever to preserve music as something that could express those special qualities, the human and the spiritual.[3] What Satie was doing seemed to be a betrayal of all this.

Yet, instead of being an aberration, Satie's innovation was really an omen. In the years following his 1920 recital, background music quickly became one of the defining sounds of the twentieth century: the sound of music as a complement to modern urban life, heard in shopping centres, cafés, offices, hotel lobbies and lifts. This, then, is a story of a sound that is banal and disposable; but it's also the story of a sound designed to get us to buy more or work faster. As we know, sound we have barely noticed can still have a profound effect on our bodies and minds. In the past this quality might have been deployed to encourage us to hunt in a more coordinated way, or feel a sense of community, or stimulate a sense of religious awe. Over the past century, music has become a carefully sculpted tool for controlling our energy and mood in the service of business.[4]

But did the creation of background music mean we handed control over the soundscape of workplaces and public space in our towns and cities – perhaps even control of our own bodies – to private interests? Or did it actually bring a little humanity and light relief into lives that were, by the 1920s, becoming rather mechanical and routine?

Inevitably, it was in the heartlands of twentieth-century capitalism that background music first took off – in some cases, quite literally. When the Empire State Building opened in New York in May 1931, music was piped into all the elevators, lobbies and observatories. One reason, according to Joseph Lanza in his history of background music, was simply that electric lifts, like airports or rollercoasters, were still novel and unsettling places. He calls them 'floating domiciles of disequilibrium': spaces in which people would naturally feel nervous about what technology was being asked to do; namely, lift them into the air at incredible speed with just a few cables and motors and bolts to keep them safe. It was realised that all fears of imminent death or feelings of motion sickness might be banished if just a few notes of calming music were heard, helping to make these unfamiliar places a little more familiar, a little more palatable.[5]

The Empire State Building was the most spectacular example of 'elevator music' in action, but it wasn't the first place in America to find a practical application for Erik Satie's notion of furniture music. The foundations for something even more radical – and certainly more profitable – had been laid a few years earlier in nearby New Jersey and Staten Island, by an ex-army radio engineer, George Owen Squier. In the mid-1920s, Squier had worked out a system for wiring canned music into restaurants, hotel dining rooms, offices and shops in return for a modest subscription fee. Grocery stores, for instance, would be supplied with a mix of music and announcements about special offers; restaurants and hotels would get a continuous flow of music. The sound would come from a central studio equipped with several record turntables working side by side, so that one disc could start spinning as soon as another was about to end. In this way, the output could continue without interruption for nearly eighteen hours a day.[6]

The company for which Squier worked was originally called Wired Radio, but as it became more and more successful, and other companies copied the technique, Squier wanted a catchier title. It was renamed Muzak, and when, in 1936, Muzak moved into bigger offices in Manhattan's Fourth Avenue, the scale and sophistication of the operation reached new heights. Customers could by now choose from four different networks of piped music: the Purple Network, specially designed for daytime restaurants; the Red Network, for 'modest sized bars and grills' wanting a mix of sports, news and weather; the Blue Network, for department stores; and the Green network, for private apartments. Muzak also started getting musicians into its own studios to record music specially adapted for its networks.

What really distinguished this specially arranged music, though, was that it was neither classical nor jazz, neither show-tune nor waltz. It was a hard-to-pigeonhole amalgam of all

these styles with the hard edges rubbed away: hot, pounding rhythms replaced by lush, gentle strings. It was 'light-
music', and, as the American composer and pianist Morton
Gould said, exactly the kind of music that would 'please the
most number of people and offend the least'.[7] This, in other
words, was perfect for a sound that was to be heard but not
listened to.

This middle-of-the-road style wasn't its only feature. By the
1940s a further refinement was introduced by the Muzak Corporation: 'Stimulus Progression', a term with an unmistakable
Big Brother ring to it, and with some justification. Individual
pieces of music were rated on their mood and pace, from, say,
'Gloomy – minus three' all the way to 'Ecstatic – plus eight'.
They were then programmed in sequence so that, over time,
the music heard in a shop or on a factory floor moved almost
imperceptibly from downbeat to upbeat in a series of fifteen-
minute segments. Over the years, further refinements crept in:
a greater concentration of cheery, caffeinated melodies over
breakfast, say, or a greater stress on classical music at dinner-
time. But the underlying principle was always the same. The
music's 'ascending curve' would give workers a pep-up just as
fatigue set in and their efficiency declined.

In effect, this was nothing more than the application in
sound of Frederick Winslow Taylor's hugely influential 'Principles of Scientific Management', which had been published
in 1911. Taylor had shown how both time and energy could
be saved if factories insisted on their workers following more
regulated, pared-down actions without deviation. This was
the theoretical basis of assembly-line production, a technique
captured on the cinema screen in its full dystopian horror by
Fritz Lang's 1927 film *Metropolis*, where workers slavishly
move lockstep in time with the insatiable machinery of a sub-
terranean city – and then subverted so wonderfully by Char-
lie Chaplin in his 1936 film *Modern Times*, where our hero's

utter failure to follow the prescribed body movements in one short factory-shift leads to inevitable chaos.[8] Piping music into workplaces on the 'Stimulus Progression' model was simply a means of keeping workers on task through sounds that matched the rhythms and movements being demanded of their bodies. It was the creation of what Joseph Lanza has called an 'optimum work womb'.[9]

It was actually a British study that proved to be the crucial breakthrough in linking work rate with the importance of sound. The Medical Research Council had found in 1937 that playing gramophone records to factory workers involved in highly repetitive work boosted their productivity. Muzak and other suppliers of background music in America quickly seized upon the findings. But it was in Britain three years later that the lessons of the research were first applied on a massive scale.

By 1940 the country was at war with Nazi Germany. Industrial efficiency in munitions and other essential supplies had become a matter of national survival. And on 23 June, just three weeks after the retreat at Dunkirk, the BBC duly responded with a new radio programme, *Music While You Work*. The BBC's aim was unambiguous. Important though factory output was to the war effort, there was no avoiding the fact that production-line work remained incredibly monotonous. Reports were coming in that workers, many of them now women, were bored, that morale was low and, most alarming of all, that concentration was lapsing towards the end of a shift.[10] *Music While You Work* was tailor-made to tackle the problem. Twice a day – for a half-hour mid-morning and then again in the afternoon – radio sets in factories across the country were turned on and workers heard a sequence of music uninterrupted by announcements, which the BBC had specially chosen for being 'rhythmical' and 'non-vocal'. As one of its producers explained, the music 'should form a background, so that listening becomes almost subconscious'.[11]

The crucial question was whether *Music While You Work* actually did what it was designed to do. One producer, who visited a factory the morning after an air raid in March 1943, had no doubt:

> I observed the tired, drawn faces, the wearied droop of
> shoulders, the glances at empty seats and felt that the very
> air was filled with the nervous tension of the past night ...
> Suddenly the loud-speakers came to life – a voice was heard
> 'calling all workers' – and then followed the rousing strains
> of a march familiar to millions of people in this war and
> the last – 'Colonel Bogey' ... Like a trumpet call to action,
> the martial melody echoed through the shop, and then I
> witnessed a transformation scene – tired faces breaking into
> smiles, the squaring of bent shoulders, chins uplifted, and
> suddenly voices, singing voices, that from a murmur swelled
> into a roar as with heads raised in defiance those factory
> workers shouted: 'AND THE SAME TO YOU!' [12]

The programme wasn't always this successful, however. To begin with, the orchestra the BBC employed to perform all the music live in its London studios kept pausing between numbers to retune their instruments, which meant that crucial momentum was lost. It also took a year or two for the BBC to find the right mix and flow of sound. By a process of trial and error it progressively tightened the selection of music, banning anything that was too complex, encouraging the orchestra to use only what it called 'moderately-loud-to-loud power', and ensuring that strict time limits were set for each tune. 'Just try to make the period one of unrelieved BRIGHTNESS and CHEERINESS,' one of the producers told the orchestra. [13]

The BBC also tried to ensure that *Music While You Worked* played tunes that factory workers actually liked. This was a significant departure from usual practice. When it came to a strict concern with productivity, it had been assumed that highly rhythmic music was always best – and that meant marching music and certain dance tunes. But the evidence that listening to the programme actually raised output was, at best, mixed. What it *did* raise was workers' morale, and that was dependent, it seemed, on them hearing music that was familiar. Indeed, the band-leader Wynford Reynolds, who helped organise the programme, believed it was vital that it did everything it could actively to avoid either music or workers being reduced to mere cogs in a war-machine:

> Workers definitely prefer tunes that they know and the most
> popular programme is one which enables them to join in
> by humming and whistling ... Music is a mental stimulant.
> It has a humanising effect which helps counteract the evil
> effects of the mechanisation of workers. This, indirectly, is a
> decided benefit to production.[14]

As the historian Christina Baade shows, *Music While You Work* turned out in the end not to be yet another instrument of Frederick Winslow Taylor's theory of industrial efficiency and scientific management; rather, it was an attempt to correct Taylorism's most dehumanising aspects. It tried to remove some of the relentless monotony and boredom of the assembly line. Indeed, for Wynford Reynolds, here surely was a noble example of music as medicine; music that retained its age-old spiritual power. This more holistic approach certainly resonated with the British public. By the end of the war not just 8 million factory workers but many millions more at home were listening to *Music While You Work*, claiming to find it a real tonic when faced with household chores.[15]

The programme worked on another level, too. The fact that both factory workers and ordinary radio listeners were tuning in meant that everyone on the Home Front felt part of the larger war effort. These were not people listening passively: when they hummed and whistled and sang along, they were weaving into being a sense of community, even a sense of egalitarianism. It was what helped make the ideal of a 'People's War' into a reality. This challenges the idea that background music is always and everywhere a malign political force acting upon and controlling our bodies or minds. We are usually too good at taking hold of sound and making it our own for that to keep happening.

And yet, if background music doesn't help someone make money, it's hard to explain why it's still so prevalent today. One reason, perhaps, is that companies such as Muzak, along with countless shops and cafés and offices, long ago learned the lessons from *Music While You Work*. They too now take the long view that even if we can't be driven to ever-faster work rates, we can at least be induced to feel a little more mellow and contented, more likely to hang around a little longer, to have another coffee, spend a little more cash. A generic middle-of-the-road light-music arranged according to strict rules of 'Stimulus Progression' simply doesn't measure up to a world of widely differing musical tastes and lifestyles. Perhaps it never did. The aim now is to provide something more carefully segmented, so in the space of wandering around a few blocks in a city such as New York, for example, we might hear classical music in a hotel lobby or a shopping mall at closing time, some middle-of-the-road rock at a chain store, and a fashionable mix of new wave or alternative folk to accompany the vaguely bookish and bohemian chatter of an East Village café. In each case, we are still dealing with music that should not be listened to, but should still be heard. In each case, the music is there to nudge our mood in certain directions. And for most of the time, this is harmless enough, perhaps even positively pleasurable. Like radio, it's a therapeutic means of feeling more relaxed, at one with the world, distracted from our more troubled thoughts, bathed, as Joseph Lanza puts it, in a comforting aural amniotic fluid.[16]

Perhaps, though, it isn't background music's effect on our mood as we sit or stand in any particular place that matters. It's the tendency it has to follow us wherever we go, squeezing out all other sounds as it does so. Even the producers of *Music While You Work* decided that the programme should be broadcast only in small doses, and never more than two-and-a-half hours a day. Such limitations have been tossed aside, and we now seem a little closer to the dystopian existence once conjured up by Aldous Huxley in his 1932 novel *Brave New World*, a world in which 'it was happiness rather than truth and beauty that mattered', where piped music of 'hyper-violin, super-cello and oboe-surrogate' filled empty spaces continuously with the 'agreeable languor' of 'zombie symphonies'.[17] The sheer ubiquity of background music has since inspired decades' worth of vitriolic reaction. J. B. Priestley once bragged about having 'had it turned off in some of the best places'. Spike Milligan later claimed that 'Tranquility is something that liberates the soul; Muzak destroys it.'[18] And, more recently, the American writer George Prochnik has suggested that it isn't even background music any more, but foreground music, attempting to grab your attention through being unashamedly loud.[19]

For these observers – and, we know, for many thousands of us – whatever the merits of an individual tune, its unshakeable presence changes everything. It makes it a noise and nothing more; a form of pollution. It's just one more unwanted sound in a world where it feels as if noise is spinning out of control on almost every front.

29

An Ever Noisier World?

In 1926 the intersection of 34th Street and Sixth Avenue in New York City was awarded a particularly dubious honour. After careful measurement by specially trained engineers, it was officially identified as the noisiest place in what was then generally regarded as the noisiest city in America.[1] It's a standard feature of global twentieth- and twenty-first-century life that cities are noisy – *all* cities – and it's likely that the din of Mumbai, Rio de Janeiro or Nairobi would now easily outstrip that of Manhattan. But in the 1920s Manhattan was the world's greatest centre of consumerism, and it had a soundscape to match: shop after shop used loudspeakers above their doors to advertise their goods to the whole street, aeroplanes would fly low overhead for hours at a time broadcasting slogans and jingles to the people below, cars and lorries

and taxis hurtled noisily along streets, and there was an endless hammering and pounding as skyscrapers were built. This
cacophony was captured in the *Saturday Review of Literature*
on 24 October 1925:

> The air belongs to the steady burr of the motor, to the
> regular clank of the elevated, and to the chitter of the steel
> drill. Underneath is the rhythmic roll over clattering tiles of
> the subway; above the drone of the airplane. The recurrent
> explosion of the internal combustion engine, and the
> rhythmic jar of bodies in rapid motion determine the tempo
> of the sound world in which we have to live.[2]

Only a few decades earlier, people had been complaining
vociferously about horse-drawn carriages, musicians, peddlers,
farm animals and church bells. The modern world had since
created a whole new battery of sounds. When New Yorkers
were asked in 1929 to list their top ten most irritating noises,
they all turned out to be machine-made: car horns, radios
blaring and, above all, the relentless roar of street traffic.

Over the past hundred years, most of us have become familiar with such noise. Most of us, too, have accepted it as irritating but also probably necessary, the unavoidable by-product
of material progress. The American author Garret Keizer captures it in the title of his book *The Unwanted Sound of Everything We Want*,[3] and this equation of noise with consumption
explains why it has been so hard to complain about it without
coming across as hopelessly nostalgic or as a killjoy, denying
fellow human beings their enjoyment of the trappings of modern life. Yet many of us have also come to believe that at one
moment or another in the past hundred years a crucial tipping
point was reached, that, thanks to capitalism, or globalisation,
or even the information revolution, we arrived at a stage in
human history when the noise of progress became self-defeat-
ing – a stage when, as the *Saturday Review* predicted back in

1925, the 'pitch of modern life' was raised beyond endurance and, perhaps, beyond repair:

> No one strolls in city streets, there is no repose in automobiles or subways, nor relaxation anywhere within the range of a throbbing that is swifter than nature. Our nervous hearts react from noise to more noise, speeding the car, hastening the rattling train, crowding in cities that rise higher and higher into an air that, far above the grosser accidents of sound, pulses with pure rhythm.[4]

Faced with this onslaught of noise, the first line of attack – at least, in a commercially thriving city such as New York in the 1920s – was to appeal to business leaders' own pockets. It was to argue that noise, far from being the healthy hum of industrial activity, might actually destroy workers' efficiency and therefore hit profits. The crucial piece of evidence came from a Manhattan office typing pool. In 1927 the industrial psychologist Donald Laird compared the typists' speeds, and the number of mistakes they made, when they were switched from a quiet to a noisy environment. He also hooked them up to machines to calculate the rate at which they burned up calories. The results were startling. The best typists worked about 7 per cent faster when it was quieter. Under noisy conditions the typists used on average 19 per cent more energy – mostly, it seemed, because their stomach muscles had tightened involuntarily, as if they were experiencing a minor form of primal fear reaction. The conclusions were obvious: noise affected workers' bodies, and noise therefore cost money.[5] People's nervous energy represented a commercial asset, perhaps even, when multiplied among millions, a precious national resource, and it was apparently being steadily eroded by the sheer din of city life.

It was evidence like this that stimulated New York's first city-wide anti-noise campaigns. Within two years the city had established a 'Noise Abatement' Commission, which put an official stamp on pioneering work by one of New York's most seasoned campaigners, a wealthy Manhattan physician, Julia Barnett Rice. She had been trying for years to limit the piercing whistles sounded night and day by tugboats on the Hudson River. She had also established a Society for the Suppression of Unnecessary Noise, and even convinced the city authorities to create quiet zones around schools and hospitals. Again, what had strengthened her campaign was that by distinguishing carefully between 'necessary' and 'unnecessary' noise she had managed to retain the support of businesses. And again, she had appealed to notions of efficiency, having shown, for example, that teaching time in schools was sometimes reduced by as much as 25 per cent whenever it was noisy outside the classroom. By 1929 the idea of measuring noise levels, and of then creating separate 'zones' for housing, businesses and industry, was broadly accepted. A few novel measures had also been introduced, such as using traffic lights rather than police officers with whistles. Yet the noise enforcers soon realised that any attempts to reduce overall noise levels were doomed to failure.[6]

One reason was simply that these attempts were often aimed at the wrong target. The first concerted effort to reduce noise levels had been at Coney Island, where in the early decades of the twentieth century New York's working-class families came every weekend in their tens of thousands to enjoy the food, the music and the riotous spectacle of the fairground. Even before the First World War, fairground barkers had been banned from using megaphones to attract passing customers. This approach was gradually extended from Coney Island to the rest of the city, so that street sellers, newspaper delivery boys, buskers, roller-skaters, even those who carelessly kicked a tin can down the street, were all targeted. This clearly satisfied the middle-class New Yorkers' vision of a well-ordered city, but it did nothing to tackle the real source of most noise: traffic. Indeed, it made the problem infinitely worse. By removing all the peddlers and buskers and playing children, it had cleared the streets so that they became mere traffic arteries rather than places of sociable human interaction.[7] There are enduring lessons here. Noise enforcers, it seems, need to be careful what they wish for.

Beyond New York, beyond America, beyond the West, there is of course a wider world that still contains a startling variety of different soundscapes, but nowhere, it seems, has the trend been towards greater peace and quiet. Among what locals call the 'salad' of sounds in the Pau da Lima favela of the Brazilian city of Salvador, for instance, the past decade has seen a new ingredient in the soundscape: the noise of new evangelical Christian churches competing for custom with older Catholic places of worship. In this battle for souls, making your presence felt – establishing one's turf – has been essential. The various churches fling their doors wide open, the singing gets louder, and sermons are relayed via loudspeakers strung up on lamp posts. When the church sermons are done, evangelical DJs take over the loudspeakers, their exhortations pulsing through the air well into the night.[8] In Ghana's capital, Accra, several evangelical Christian churches and even one or two Islamic mosques have recently been criticised by residents for their excessive noise-making. In one instance, people who live near a gospel church have claimed that it has become impossible to talk or make phone calls in their own homes every time a service is held; and one resident has even claimed that when the pastor was asked to turn down his public-address system, he simply responded by asking his congregation to shout and clap even more loudly.[9] The city authorities have tried to act, but so far to no avail. In the meantime, it would appear that as long as religious fervour is on the rise – as it appears to be in many parts of the world – religion itself is going to keep getting louder.

Nevertheless, when it comes to the defining sound of the late twentieth century, God has surely been outdone by Mammon. The consumerism that characterised 1920s Manhattan has long since gone global. In every corner of the world, there are now people who want more things, better things, newer things – and louder things. Above all, almost everywhere, people want a car, with the result that on any main street in almost any city or town you choose to visit it is now next to impossible to sustain normal conversation for any length of time. Western travellers who arrive at a city such as Accra, for example, frequently describe the visceral shock as they experience for the first time the explosion of sounds from traffic, car horns and blaring loudspeakers. The degree of noise is extraordinary. But really, it's no more than an exaggerated version of a broader trend: as background levels of sound have risen, we have all tried to mask the roar with other noises of our own making, and we have all been driven to greater noise simply in order to be heard.

In the quieter parts of the Ghanaian capital, there is another dimension of globalised twentieth-century noise: a creeping sense of sameness. There are plenty of shops in downtown Accra blaring out distinctly Ghanaian sounds: the highlife and hiplife noted previously, even some traditional Ashanti music. But then there are the countless hotel lobbies playing western classical tunes, or at least ersatz versions of them. As for the city's international airport, with its burbled announcements, the steady tread of feet on linoleum floors, the squeak of luggage trolleys, the piped music, it sounds very much like any other international airport. The literary historian Steven Connor refers to this kind of atmosphere as a 'murmuring emulsion of sound, compounded of all the sounds of rush and hurry, mechanical and human'.[10] An airport is the purest example of a place in which sounds have become so thoroughly homogenised that it ceases to be an individual place at all: what reaches our ears gives us no clue to where we are. For many of us, of course, the standardised soundscape of international travel is one of the things that helps makes it so bearable: there is always comfort in the familiar.

However, the amplified noise of commerce and traffic and travel has undoubtedly used up a lot of the available oxygen that once allowed quieter 'indigenous' sounds – either human or natural – to breathe comfortably and make their presence felt. As Garret Keizer points out, the sound of human conversation is a good match for a rainforest, but not for a jet ski or a chainsaw.[11] In the Arctic Circle, one of the most ubiquitous sounds is now the snowmobile – so much so that not only have rates of hearing loss among native people risen dramatically, but other sounds – conversation, traditional Inuit songs, the sounds of arctic wildlife – have had to go unperformed or unnoticed. Even more worryingly, just as people living in the most overcrowded and noisiest parts of cities have markedly poorer health than their richer counterparts, animals regularly exposed to snowmobile noise, such as elk and wolves, have exhibited signs of weakened immune systems. In the world's oceans, meanwhile, whales, which depend on great acoustic sensitivity to communicate with each other, have become disorientated and distressed as a result of the sonic effluence from shipping.[12]

Disturbing revelations like this have led many people, especially since the 1960s, to think of noise as the sound of excess – a form of pollution, in effect – and indigenous sounds as being like species under threat of extinction. The musician and naturalist Bernie Krause is one of the most eloquent voices in this environmental debate. Krause has devoted much of the past four decades to travelling the world and recording wild soundscapes. He has amassed something like 4,000 hours of material, but calculates that, due to encroaching human activity and noise, a good half of the places he has visited over the years no longer survive as distinct and unpolluted sonic environments. He hopes that by drawing people's attention to the beauty in natural sounds he might change human behaviour; yet his work also has an uncomfortably elegiac feel to it. It's as if we have reached a point where the untouched natural soundscape will only survive as a carefully preserved museum exhibit, to be listened to with curiosity and wonder by future generations. In the meantime, Krause suggests, amplified noise in all its forms diminishes tragically the range of experience on offer to humanity.[13] Which is why Garret Keizer, despite the title of his own book, wonders whether we now need to think of noise not as something *unwanted* but as something *unsustainable*.[14]

Noise, then, has become the sound of excess, although it is no longer just the sound of excessive materialism. In the Internet age, noise is also the best metaphor for the excessive quantities of information with which we are faced on a daily basis. We talk of the 'static' in our lives, the 'noise machine' of political rhetoric, the need to distinguish between 'the signal' and 'the noise'. In each case, noise stands for the disruptive fog of junk data. Like spam emails, it gets in the way of good data. Worse, it prevents us from hearing ourselves think.[15] It is, as Steven Connor puts it, 'irrelevance, distraction, indigestible, disorganised'. And, perhaps because it is so closely allied with all the machinery and electronic devices that fill the world with automated signals – from the ping of alerts on our mobiles and tablets to the wail of car alarms – it sometimes feels as if it's grown a mind of its own.[16] In practical terms, noise that *seems* unstoppable like this is as dangerous as noise that really *is* unstoppable. After all, it's always other people's sounds that cause anxiety and stress, not our own – and that is because theirs is the sound we cannot switch on or off at will. That is precisely what makes it 'noise', as opposed to mere sound.

So if noise is unstoppable, might it at least be manageable?

Back in New York before the Second World War, there were still a few creative, warm-hearted souls who were willing and able to face up to the cacophony of city life by embracing it. One was the musician, Duke Ellington. In his classic tune 'Harlem Air Shaft', we catch, drifting up and down the ventilation shafts running through tenement buildings, linking one apartment to another, the unavoidable sound of humanity, with all its faults and virtues. As Ellington explained:

> So much goes on in a Harlem air shaft ... You get the full essence of Harlem in an air shaft. You hear fights, you smell dinner, you hear people making love. You hear intimate gossip floating down. You hear the radio. An air shaft is one great big loudspeaker.[17]

Of course, one reason Ellington could bear the thought of all this being amplified was that it was mostly the sound of fellow African Americans: people whose voices had been kept silent for too long. It was sound that also spoke of a shared experience, a sense of community; sound, in short, which needed advertising to a wider audience. As the Harlem poet Langston Hughes wrote, 'the blare of Negro jazz bands and the bellowing voice of Bessie Smith singing Blues' was something that now *demanded* to be heard as a matter of ethics as much as aesthetics.[18]

In the twenty-first century, when so much noise is machine-made, amplified beyond endurance and behaving as if it had a life of its own, when even the sound coming down an air shaft is from anonymous strangers, not well-known neighbours, can it still be embraced in this generous, almost celebratory fashion? It seems unlikely. Indeed, it appears there is a growing desire to turn our backs on noise, and sometimes on our fellow human beings, in pursuit once again of that rare and elusive quality, silence.

30

The Search for Silence

Three times a year at the Ashmolean Museum in Oxford there is an intriguing opportunity for the jaded visitor to step away from the stunning displays of antiquities, as well as the hundreds of other visitors strolling and chatting as they wander around, and experience what can only be described as a Zen-like moment of peace and tranquillity. It's to be found not in the rooftop or basement cafés – which, though pleasant enough, are usually loud with human chatter and the clinking of crockery – but in a small gallery tucked away on the second floor. It's here in Room 36 that you will find the Ningendō Tea House, a delicate bamboo structure where on certain pre-advertised dates you can witness, or even take part in, a Japanese tea ceremony.

As a guest at one of these ceremonies, you are required to

enter the Ningendō Tea House through a small, low opening. It forces you to bow down in a posture of humility and
reverence, much as you have to when making your way into
the Maeshowe chambered cairn on Orkney. And once again
you find yourself entering a different realm. The interior of the
tea house is stripped of ornamentation; it is clean, calm and
harmonious. As you kneel on a mat, the host enters through
another opening at the back and begins a highly orchestrated
set of manoeuvres. Aside from a few brief exchanges, the two
of you hardly talk. Amid the silence, the tea is prepared, your
ears stimulated at each stage by the subtlest of sounds: the
host, her silk kimono whispering as she moves, ritually cleans
the utensils, pours hissing water from the kettle into a bowl,
uses a piece of bamboo to whisk the tea into a bubbling froth
and then passes it to you. You are required to take your time,
turning the bowl around in your hands, putting it down to
admire it before picking it up again, sipping gently, bowing,
giving quiet thanks. For a cup of tea, it's quite a palaver. But
then, as the observer of Japanese culture Lafcadio Hearn
wrote back in 1905, what you are experiencing is not simply a
quenching of thirst but an art based on years of training: 'The
supremely important matter is that the act be performed in
the most perfect, most polite, most graceful, most charming
manner possible.'[1]

In Japan itself, this whole process might take several hours.
You would probably first approach the tea house by winding
your way slowly through a garden of trees and bushes, moss
and dew – the sound of the city receding into the background,
your focus turning inwards – before washing your hands at a
basin as an act of purification and then removing your shoes.
The ceremony itself would be more elaborate too, involving
a greater number of ritualised movements and exchanges.
But even in the Ashmolean Museum, with the clock ticking
towards closing time, people sitting nearby watching and a

constant background murmur coming from neighbouring galleries, the Ningendō Tea House still manages to conjure up a sense of nature, as if the trickling, bubbling and hissing you hear were really the sounds of the wind whistling through the pine trees of a Tokyo garden. As the host Kyoko Regan explained to me after I had taken part in a tea ceremony here, the whole point is to tell stories without speaking, and, above all, to 'create an atmosphere'. Not just any atmosphere, of course: it is very consciously one in which we are able to 'grow ourselves inside'. For a precious hour, she carves out a space, not quite of total silence, but certainly as close to silence as might be possible on a working day in a busy city. In short, she provides a sonic oasis for contemplation.

It's no coincidence that tickets for the Ashmolean tea ceremonies sell out fast, nor that these ceremonies are now hugely popular events at museums all over the world. The tea ceremony is only one among a smorgasbord of highly popular sensory experiences for sale. In the rolling farmland of Iowa, for instance, it's possible to book yourself into a three-day silent retreat with the Trappist monks of New Melleray Abbey. Visitors to the abbey are politely advised to leave their portable radios, CD players, and so on behind so they can escape 'the bustle of business, the chaos of the streets … the blare of the media' and 'the mad rush and noise of modern life'.[2] The point of a retreat, they are told, is that through silence and solitude the opportunity might be found for 'bumping into yourself'[3] – though, as it happens, the opportunity may prove rather elusive: rooms at New Melleray Abbey are usually solidly booked up for further ahead even than places at an Ashmolean tea ceremony. It seems that in the popularity stakes monastic retreats now outdo even wilderness camps, or the chance to spend a long weekend up in the mountains howling with wolves. And if all these options for getting away are simply unavailable to you, there remains the cheapest and easiest option of all: to stay at home and read; in particular, perhaps, to lose yourself in one of the burgeoning number of books about silence that have been published in recent years: George Prochnik's *In Pursuit of Silence*, or Sara Maitland's *A Book of Silence*, or even John Lane's *The Spirit of Silence*. The list of available titles could go on; however, the point, I hope, is obvious by now: in a world full of noise, the value of silence is rocketing.

We can see all this as the equivalent in sound of all those Slow Food or Slow City movements which took off in the 1980s and 1990s. Like them, the search for silence represents a laudable effort to halt in its tracks the bustle and rush of consumption. It seeks, as Prochnik says, to 'find something positive' in the ways of life that flourish when things are stiller or quieter.[4]

Yet it's also possible to view the world of Zen-like tea ceremonies and three-day monastic retreats with a more sceptical eye. Garret Keizer points out the cruel irony of westerners jetting off to silent retreats in India, shattering the quiet of anyone living below the flight path. He also reminds us that whenever we sit quietly reading a book, we hold in our hands an object that has generated huge amounts of noise in its manufacture, noise that is experienced by *other* people in *other* places: those working as loggers in the forest or living near a distant paper-mill or next to a distribution depot rumbling with delivery lorries.[5] So none of these moments of personally replenishing silence actually solves the problem of pervasive noise 'out there'. Indeed, they sometimes make it worse. Their problem is that they are private solutions, like turning our back to the world, pulling up the drawbridge and telling everyone else, in effect, to go hang. Indeed, it is this age-old attempt to flee a noise rather than tackle it at source which keeps coming back to haunt us in this history of sound. The elites of ancient Rome fled to their villas on the Palatine Hill;

the wealthy of eighteenth-century Edinburgh built a whole New Town for themselves; Thomas Carlyle created a window-less study in his attic. And in the past hundred years, before we had got round to reviving all these silent retreats and tea ceremonies, we tried to cut ourselves off from the sounds of the streets by deploying the very latest discoveries in science and engineering.

One early example of this occurred on the eve of the First World War with the reinvention of the humble tile. St Thomas Church on Fifth Avenue, New York, might look to the casual observer just like one of medieval Europe's great Gothic stone cathedrals, yet it was built only in 1913, by which time it was technically possible to correct that irritating acoustic flaw of the Gothic interior, namely the reverberation that creates beautiful music and makes sermons so difficult to hear clearly. The inner surface of St Thomas's vaulting was covered with brand-new tiles. Each one was made up of a special new mix-ture of clay and earth and had lots of tiny interconnected air spaces inside to act as bafflers. Buildings had been insulated from outside noise for years, their walls filled with padded paper, or grass, or cattle hair. Most of this, however, had been ad hoc stuff, added hastily afterwards like a sticking plaster. St Thomas was different because these tiles – so-called 'Rum-ford' tiles – had reverberation-reducing materials engineered into their very fabric right from the beginning.[6] The acoustic results were a source of inspiration for a new generation of architects and engineers, who now dreamed of creating all sorts of buildings entirely soundproofed from the din of Man-hattan's streets.

Their efforts were realised in 1928 with the opening of a building on Madison Avenue that would house the new head-quarters for the New York Life Insurance Company. As the acoustic historian Emily Thompson has shown, Cass Gilbert's design incorporated every technique then known for reducing sound. The walls were filled from top to bottom with a special felted material made of sanitised cattle hair and asbestos. Windows were made of extra-thick glass in heavy frames, and none of them ever needed to be opened because there was artificial ventilation. All the plumbing and machinery operating the high-speed elevators or the pneumatic tubing for delivering mail was located well away from office areas; as were the kitchens, which delivered food to canteens via special belts and lifts. In the offices themselves, as well as in all the corridors, only the most effective soundproofing materials were used: partitions were made of heavy metal and glass; ceilings were covered with acoustic tiles; floors were covered in cork to minimise the sound of footsteps.[7] The results, as *Scientific American* reported in 1929, were phenomenal:

> Imagine, if you can, a large office with typewriters and adding machines clicking away, telephones ringing, filing cabinets being opened and closed, doors shutting, clerks coming and going – but with not a sound above a murmur reaching the ears. Even the sound of the steel worker riveting outside is subdued. Such a condition, which seems almost unbelievable at first, is actually typical of the work rooms of the building, and is made possible only by an extensive installation of sound-absorbing materials ...[8]

None of this came cheap, but for the 6,000 workers inside its walls, the New York Life Insurance Building was a marvel: a city-within-a-city – hushed, efficient and hermetically sealed off from all the bustling, turbulent life beyond its walls.

What, though, have been the longer-term consequences of engineering triumphs like this? Often, things haven't worked out quite as expected. Sometimes, as Emily Thompson points out, we start by seeking silence and end up discovering that silence itself doesn't actually have much character, that the hush of a modern office quickly becomes just another homogenised global soundscape. Sometimes we discover that the 'cleaner' the sound we create with modern technology, the colder and less human it feels. This, no doubt, is why many music fans now prefer listening to the analogue charms of a fuzzy, scratchy vinyl disc to the clinically digital sound of a CD or MP3. Above all, perhaps, we discover that noise isn't really a technical problem, demanding an easy (though usually expensive) technical fix; it's a social problem, demanding some tricky social solutions.

Soundproofing generally makes silence a private commodity, available to big companies or wealthy individuals rather than those who, living in the noisiest and most overcrowded conditions, probably need it most. Worse, it means that those of us safely sequestered behind soundproofed walls grow so accustomed to the quiet that we become overly sensitive to noises outside, which suddenly seem so much louder and more threatening than before. As Garret Keizer argues, the sounds of people we know and like 'seldom strike our ears as "unwanted".' [9] But when we soundproof ourselves, the people outside are going to become strangers to us. Having cut ourselves off from this wider world, we miss all sorts of sounds that we really ought to hear. This acquired deafness has been captured brilliantly in the recent television drama series *Mad Men* – set, as it happens, on Madison Avenue, the same avenue on which Cass Gilbert's magnificently soundproofed Life Insurance Building stands. *Mad Men*'s cast of characters spend their working days in a plush, hushed ad-agency office, trying desperately to figure out the latest twists and turns of public taste but struggling – and failing – to understand the dramatic changes in American society wrought by feminism and the civil rights movement. The ad-men's misunderstanding is the price they pay for being unable to *hear* the 1960s hurtling towards them from Harlem and Greenwich Village.

As George Prochnik says, 'Soundproofing is terrific like

bulletproof flak jackets are terrific.' But, he adds, 'wouldn't it be better still if we didn't have to worry about getting shot all the time?'[10] The answer has to be a resounding yes, for noise can only be successfully addressed if we engage with it in the public arena as a whole.

When we have done this in the recent past, the change has sometimes been dramatic. In Dutch cities such as Amsterdam, Utrecht or Maastricht, for instance, pedestrians and bicycles have long been given priority over motor traffic as a matter of policy. As a result the soundscape in their streets and piazzas and arcades is strikingly different from Britain's car-ravaged town centres. Far from being deafeningly loud – or, indeed, totally silent – these Dutch cities provide their citizens with a vivid spectrum of sounds that have been smothered or extinguished elsewhere: street vendors, footsteps on cobbles, church clocks and bells, conversation, laughter.[11] For those of us living in less progressive cultures, acoustic interventions have had to be on a more modest scale. Architects and town planners, for instance, have tried creating 'pocket parks' – small patches of land just large enough to provide a resting place set back from the traffic. In New York, the original idea of the pocket park is credited to Jacob Riis, who, as we learned in Chapter 24, campaigned hard for better living conditions among the tenements of Manhattan's Lower East Side. But the first one – Paley Park, just off busy Fifth Avenue – wasn't actually opened until 1967.[12] In Britain, it has taken even longer for cars to be blocked from the north side of Trafalgar Square, so that people sitting on the steps of the National Gallery can enjoy the sound of the nearby fountains.[13]

What unites all these examples, whether in New York, London or Utrecht, is that reining in the oppressive noises of industry or traffic has not put silence in their place instead. This is not only because it would be unrealistic to do so; it's also because we don't actually *like* it. Total silence sets our subconscious alarm bells ringing. It's why the famous film-editor and sound-designer Walter Murch so often uses moments of pure silence in his soundtracks. In *Apocalypse Now*, for instance, Murch slowly drains all the sound from a scene set in the jungle. Without really knowing why, both the viewer and the characters on screen are put on edge. We sense that something is wrong – which, indeed, there is, since the silence, we soon discover, marks the presence close by of a prowling tiger.[14]

Silence, then, can be as overpowering as loud noise, and our deeply ingrained unease towards it is precisely why many of those workspaces so carefully soundproofed in the first half of the twentieth century have since been refilled with artificial noise – like one in London, which in 1999 ordered a special machine to be installed to play background 'mutter' after the accountants working there complained of stress, and even feelings of loneliness, caused by too much quiet.[15]

The author of the Harry Potter novels, J. K. Rowling, explained how she did most of her writing in a café in Edinburgh. Apart from anything else, she said, being in a café kept her warm. But warmth is more than a matter of temperature. It's also about a convivial atmosphere, and, perhaps subconsciously, a certain amount of noise. In 2012 a Canadian study asked a group of university students to compose bits of creative writing in a range of ambient soundscapes. The most productive setting, especially for the most creative students, turned out to be a busy café with a steady amount of background chatter; somewhere, in other words, which offered a wallpaper of human sound.[16] No doubt part of the reason for the result shown by this study is that the sound itself is useful at a basic utilitarian level, simply because it conceals any sudden noises that might startle or distract while we are trying to concentrate. But part of it too, surely, is the cultural impact of the sound, that at a semi-conscious level it connects us directly with other living beings.

Sound does that, and has done throughout human history, for good or ill. Our search for silence is really a search for the space to think rather than a desire for the silence itself. And, anyway, it turns out in the end that we think best, and probably treat each other best, whenever we can hear each other close by. In coming to terms with a noisy world, equity, moderation and a human scale are all vital tools in our armoury. The very nature of sound means it is always to some degree common property. As it moves freely through the air, it demands a certain live-and-let-live ethos. The Dutch had a slogan for it during a noise-reduction campaign back in the 1970s, and maybe it's time to revive it. It simply said, 'Let's be gentle with each other.' [17] That might sound a little wishy-washy to contemporary ears, bruised and bloodied as they are by all our disputes and anxieties and suspicions. Yet it reminds us that even today sound has to be managed not by technology or by force but by ethics. It requires a world where none of us is noticeably louder than anyone else, and where none of us is cowed into deathly silence, but where all of us can hum and whistle and talk to each other – and hear others doing the same – as we go about our daily lives.

Epilogue

The 2011 film *Perfect Sense* imagines a virus sweeping the world and robbing people of their senses one at a time. It's a story that can only end in darkness and silence. But all is not lost for our characters on screen – at least, not to begin with. They start to remember how resilient, how adaptable, how wonderfully creative people can be. So they adjust to a new social order. When they lose the ability to smell or taste, our heroes rediscover the intense pleasures to be found in touch; street performers put on shows in which smells are communicated through sound; instead of creating interesting flavours, restaurants concentrate on creating fabulous textures. But while these strange twists turn out to be bearable, and indeed briefly bring people closer together, it's when the film's characters lose their sense of hearing that we see the wider social fabric begin to fray catastrophically. Communication breaks down; feelings, opinions and information go awry or get trapped inside people's heads; misunderstandings multiply; horizons narrow. The point at which people no longer hear each other is the pivotal moment of the film. It's when we as viewers sense that human ingenuity will no longer be enough, and that, with everyone's world shrinking drastically, oblivion is imminent.

Daft the film may be. But it's a symbolic reminder of how deeply *social* the sense of hearing is in our lives even today. And it's also a reminder of what we miss in our understanding of history when we don't listen to the sounds of the past. The historian Mark M. Smith points out that if we were to watch almost any film or television programme with the volume on mute, we would miss a whole layer of texture, meaning, narrative and, perhaps especially, emotion.[1] Obviously, without seeing the film or programme, we would lose a great deal, too. But the visual action is explained – 'anchored', we might say – through the soundtrack. So much of the basic storytelling information remains verbal. More than that, the feelings, and therefore the motivations, of its protagonists are often revealed most potently by the music or sound effects rather than by what can be seen in the frame. It's as if hearing takes us beyond the surface of things and allows us access to other people's minds. In our own past, similarly, it has been sound and our ability to interpret its subtle meanings that has, for good or ill, both helped us manage our moods and connected us with other people. It's why knowing the world through our ears is, and always has been, different from knowing the world purely through our eyes. It offers us a more immersive understanding of both the subjective and the social dimensions of past lives.

So when we turn up the volume of history and catch these sounds of the past, what do we hear? It's hard to refute the idea that the world has become a progressively noisier place. Objectively speaking, there are more people, more machines, more traffic, more media – more of everything really. And, no doubt, if we could somehow assemble a reliable timeline of background noise measured in decibels, it would be theoretically possible to demonstrate that most countries have become 'louder' over time. But we must never underestimate all the noise our ancestors would have experienced. Caves and Neolithic chambers probably reverberated with chanting and drumming; ancient cities such as Ur and Rome were crowded, uproarious and busy twenty-four hours a day; medieval monasteries resounded with bells; the ears of indigenous peoples were assaulted by drums, trumpets, bells and gunfire unleashed by European colonists; war has always been deafening for those right in the thick of it; many workers in the mills and factories of the Industrial Revolution had their hearing permanently damaged. The list could go on. As could a list of measures taken to *reduce* noise over time: thicker walls and soundproofing; replacing cobbles and iron-rimmed wheels with asphalt and rubber tyres; legislating to reduce workplace noise; regulating the amount of traffic and building work at night-time and weekends; restricting the use of loudspeakers. Indeed, given our technological ingenuity, there is no reason to assume the world is getting irretrievably noisier, and over the past century a great deal of civic effort and progressive social legislation has gone into reducing the worst offences. History, so far, has not presented us with a smooth and uninterrupted timeline taking us from a prelapsarian idyll of quietude to a contemporary cacophonous hell.

But in any case, even discussing noise in this objective fashion is a mistake. Noise, as I hope I have shown, is largely a matter of subjective experience: one person's irritating din is another person's sweet music. Which is why I have approached noise not as a separate category from all other sounds, but more as a rhetorical device for getting to the heart of sound's *social* role – and, specifically, the human dramas that revolve around sound at various points in history. If noise is, as most definitions would have it, an 'unwanted' sound, then to understand its impact on lived experience properly we need to work out *who exactly* considered a given sound as wanted and *who exactly* considered it unwanted in any particular time and place – and *why*.[2] When we look at the past in the broadest sense, even if only through a series of snapshots, the answer to these questions seems to keep coming back to a potent mix of three interwoven things: power, control and anxiety.

Of these three, power is the most brutally straightforward. Those in positions of strength and authority – civic rulers, religious authorities, well-armed colonists, slave owners, factory managers – have by and large been able to impose their standards of behaviour on those with no authority: citizens, parishioners, indigenous peoples, slaves, employees and the rest. They have decided who speaks and who doesn't, who makes a noise and who doesn't, and who has to listen to whom. In short, they have exercised their power to shape the soundscape. There is nothing surprising about this, of course. But it's worth pointing out that exploring these human relationships through sound – hearing how people were summoned by bells, subdued by guns or drums, made to sing or remain silent, exposed to monstrous levels of machine noise day and night – surely gives us an extra insight into how it felt to be powerless or poor, reminding us that powerlessness was not some abstract condition but something experienced all the time by people through their own senses. It should also remind us that when we come across any dispute over sound, and we try to work out who is being 'noisy' and who is insisting on other people being silenced, we really need to look for the power struggle being played out in the background.

Intimately related to power is the issue of control. By this

I mean that degree of control any group of people have had over the soundscape in which they found themselves living at any given time. Of course, sound, as I have argued throughout this book, is a capricious force: moving freely through the air, it has never been fully owned or manipulated by one institution or group of people rather than another, as if it were their exclusive property. The story of slaves finding creative ways to perform their own musical and oral traditions even in the face of outright oppression, or the story of French revolutionaries in 1789 using protest songs to articulate their disgust with the Bourbons, provides evidence enough of that. Indeed, we could go further and conclude that, since it's usually impossible to segregate the airwaves, the various sounds floating through them have something of an intrinsically revolutionary quality: any noise made by one group of people 'leaks' out to other groups of people within earshot, so that very often it's through sound that we discover and come to understand other cultures. This happened among the assorted mystery religions of ancient Rome as they listened to and freely borrowed rituals and traditions from each other, and it happened again nearly 2,000 years later when, in the 1930s, 1940s and 1950s, white Americans found themselves hearing African American jazz and blues music playing over the loudspeakers in their local stores, or drifting through open windows in their neighbourhood, or coming out of their radio sets in the living room. At various points in between, sound has become a kind of trading zone for human culture. The visual and literate world tends to offer a series of fixed landscapes and boundaries. Soundscapes offer something more fluid: they shift their size and shape and character moment by moment; they overlap with each other; they leach into each other in unpredictable ways.

The question of control doesn't go away, it just keeps changing its tune. One feature of sound that has been reasonably constant, however, is that people have generally been more tolerant of noisy environments whenever they have felt they had the option of reaching for the off switch. And throughout history the powerful have been able to reach for the off switch more easily than anyone else. The ruling elite of ancient Rome and the middle classes of eighteenth-century Edinburgh – indeed, the wealthy everywhere – were able to flee the noisy bustle of city centres to find the quiet and privacy they desired, and twentieth-century Manhattan office workers could insulate themselves from the roar of the streets outside through generous use of the most expensive soundproofing materials. The converse of this is that the poor and the powerless have had rather less say about the sounds to which they have been exposed. Just as they are nowadays more likely to be stuck living near a busy airport, so in previous centuries they would have found themselves living hard up against the hammering of blacksmiths, tanneries and other workshops.

So, even though we might say that one person's din is another person's sweet music, we need to remember that the world's supply of unwanted sound has generally been distributed very unevenly. The poor have always needed their sleep as much as the rich – perhaps more so. But for most of the past they have had much less chance of getting it, and have invariably paid the price. One hundred and thirty years ago, it was inside the dingy tenements of the Lower East Side in New York and among Glasgow's boilermakers that the scale of mental illness and deafness was at its most forbidding. Today, it's people living in the most overcrowded – and therefore the noisiest – parts of a city who tend to consume more prescription drugs and have lower educational attainments than anyone else.[3] Given the levels of population and economic growth now under way in countries such as China, India and Brazil, this pattern is likely to be repeated, not so much in the world's post-industrial North and West, as in the rapidly industrialising, and under-regulated, South and East. But everywhere has its pockets of overcrowding, poverty and social neglect. So if we really insist on looking for a grand pattern in the story of noise, we should look for it not so much in terms of rising volume levels but rather in terms of the growing inequalities in people's access to quiet. And in the end we should take these growing inequalities seriously because the evidence suggests that where we perceive noise to be more evenly distributed we become more tolerant and accepting of it – and, crucially, less quarrelsome with each other. The noisy bustle of a street market in, say, Accra or Istanbul is tolerable precisely because its sound is generated collectively; it is the accumulated sound of everyone in the street equally, not the outpouring of one dominant sound-maker. The sound, so to speak, is of 'us', not 'them'. Wherever a culture elevates the value of private space over public space, by contrast, complaints about noise are noticeably higher.[4]

This brings me to the third ingredient in our understanding of sound's social role in history: anxiety. What makes us anxious? Usually it's the alien, the unfamiliar, the other. And the very notion of an 'us' and a 'them' has been reinforced at key moments in the past by repeatedly distinguishing between the sounds 'we' make and the noise 'they' make – whoever 'we' and 'they' may be. On one side of this divide, sounds have been treated as pure, noble, valuable, enriching, replenishing. On the other they have been treated as alien, barbaric, savage, demeaning or outright malevolent – in short, as that apparently meaningless amalgam, noise. It's a pattern we saw in America in the seventeenth and eighteenth centuries, when European colonists dismissed the spoken languages and musical traditions of native peoples as a 'hellish' din. We saw it again in the streets of Victorian London, where writers such as Thomas Carlyle railed against the 'vile' Italian organ players outside his house 'grinding' out their valueless tunes. And we see it today, when we find a British newspaper columnist bemoaning a rail journey in which he has been forced to endure 'the man with his hissing iPod, the toddler with her electronic game or the ponderous girl from the buffet taking you through the full list of hot and cold beverages'.[5] In this diatribe, one senses that the sounds have become inseparable from the people making them. The sounds are not just abstract irritations; they are the harbingers of an irritatingly close and irritatingly un-shut-up-able contemporary mass culture. One doesn't want to hear the sounds, because one doesn't really want to notice the people.

If this is the case, here perhaps is an example of 'social deafness', in which we try to soundproof ourselves from aspects of society of which we disapprove or simply don't understand.[6] In *Chavs: The Demonization of the Working Class*, Owen Jones writes eloquently of how Britain's middle class deals with its distrust of working-class people (and, indeed, its discomfort at talking about class at all) by attacking such people's grosser cultural habits – by treating 'them', in effect, as a kind of alien and primitive 'tribe'. Sound plays a part in this. How members of the tribe talk, the music they listen to, the noises they make: these are key attributes of some alleged group identity just as much as the clothes worn and food eaten. And, as Jones would no doubt argue, these sounds are treated by 'us' with the same superficial disdain that we apply to other aspects of the tribe's lifestyle.[7] This represents a form of social deafness because we are failing to listen carefully enough to unfamiliar sounds; then, in failing to listen to them properly, we end up capable only of dismissing them as meaningless. That is why sounds that are deemed alien are invariably considered noise even when the decibel level is low.[8]

History warns us that this tendency to categorise unfamiliar sounds in this way comes at a high price. When those Europeans who colonised America in the sixteenth century or who explored the Australian outback in the nineteenth century started lumping together all the complex languages and varied musical traditions of the indigenous peoples they met into some amorphous category of 'savage' and meaningless sounds, it was a convenient prelude to denying the people themselves any claim to equal rights, even over their own land. Similarly, we might conclude that dismissing the sounds of ordinary people whiling away their time on a train comes perilously close to a demonisation of working-class culture in general – and, more dangerously still, might act as a prelude to a wider political rubbishing of certain people's right to occupy public space, or indeed to have their voice heard in the broadest sense.

Anxiety underpins all this, because whenever we withdraw into separate soundscapes – either by soundproofing ourselves from other people, or discouraging other people from making their presence felt – we make strangers of each other. And strangers make us anxious. So we pull up the drawbridge even further, becoming ever more deaf to what these strangers might be saying. In other words, dismissing unfamiliar or difficult sounds may well lead to what Joanna Bourke in her history of fear calls a 'psychological disposition' towards suspicion, even hate.[9] By warning of the potentially dire consequences of social deafness, a history of sound urges us towards greater mutual comprehension.

There is no reason to suppose we are incapable of thinking positively with our ears. For tens of thousands of years they have kept us fully alert to the world around us, helped us navigate and keep track of the hours, allowed us to forge social bonds, shaped our spiritual and cultural experiences, and simply kept us entertained. Throughout all this, listening has remained important because it has remained a profoundly active, skilful, ethical act. It's what allows us to unravel what we might otherwise dismiss as meaningless noise, and appreciate the dramatic tangle of social relationships it really contains. The sounds we humans have made through history – the most disconcerting and ugly of sounds, as well as the most pleasurable – have always been laden in meaning. They gave our ancestors – just as they give us still – a sense of place and time, a sense of danger and comfort, a sense of connection with other people. They have helped make us human. So instead of obsessing about sound in some abstract way, we need to recover a feeling for how important it has always been to daily life – and just what a great deal has been at stake in that apparently simple phenomenon, the soundscape.

Acknowledgements

Never has a book I have written been such a collaborative effort as this one turned out to be. For instance, without the radio series there would have been no book. So my first thanks are due to Tony Phillips at the BBC for commissioning the series for Radio 4, and to the formidably experienced Matt Thompson at Rockethouse for producing it. Cathy Fitzgerald also deserves huge thanks for acting so efficiently as a location researcher and additional producer for Rockethouse, as do Dinah Bird, Joe Acheson and Cherry Cookson.

The British Library Sound Archive was also a key collaborator in the whole *Noise* project, providing not just expert guidance but also invaluable access to its extraordinary collection of recordings. So I wish to thank Richard Ranft, the Head of Sound and Vision at the British Library, and his team of curators, including Cheryl Tipp, Steve Cleary, Paul Wilson and Janet Topp-Fargion. We hope that all the original field recordings Matt Thompson and I have made as part of this project will be deposited at the British Library Sound Archives, so they can be accessible to members of the public in perpetuity. Access to archive material, early printed books, recordings and valuable background information during my research has also come from: the staff of the Upper Reading Room, Bodleian Library, University of Oxford; Dr Noel Lobley at the Pitt Rivers Museum in Oxford; Davia Nelson and Nikki Silva, aka the Kitchen Sisters, who provided access to recordings held by the 9–11 Sonic Memorial Project; Dr Jacqueline Cogdell DjeDje, the UCLA Herb Alpert School of Music, University of California; Ian Rawes, the London Sound Survey; and Brian Reynolds.

In the making of the series and the writing of this book, only one organisation refused to give access to an important historic location for purposes of recording sound: the Dominicans of the Basilica San Clemente in Rome. But access and generous help was forthcoming from everyone else, including: Aaron Bebe; the Bell Ringers of St John the Baprist Church, Little Missenden, England; Lisa Budge; Michael Bradley in Kirkwall; the Choir of St John's College, University of Cambridge; François de La Varende at the caves of Arcy-sur-Cure; the Department of Psychology, Harvard University; Kira Garcia and the Lower East Side Tenement Museum, New York; Dr Caroline Goodson, Birkbeck College, London; Historic Scotland; Dominique Le Conte; Sarah Naomi Lee; McLeod Plantations, Charleston, S. Carolina; the National Museums of Scotland; the National Trust; the National Trust for Scotland; The New York Life Insurance Building; Matthew Owens, Organist and Master of the Choristers of Wells Cathedral; the Milman Parry Collection at the Widener Library, Harvard University; Marta Perrotta; Iégor Reznikoff; Sebastian Schmidt; Alan Tavener, Creative Director, Cappella Nova; Rachel Tavernor at the Reframing Activism Blog at the University of Sussex; the University of St Andrews, Scotland; and Yale University Press.

Several people very kindly agreed to read early drafts of various chapters, and offered extremely useful comments. These included: Professor Mary Beard, Newnham College, University of Cambridge; Dr Francesco Benozzo, University of Bologna; Professor James W. Ermatinger, University of Illinois at Springfield; Professor Robert Gildea, Worcester College, University of Oxford; Professor Chris Given-Wilson, University of St Andrews; Professor David D. Hall, Harvard University; Dr Jowita Kramer, Oriental Institute, University of Oxford; Joseph Lanza; Dr Noel Lobley, Pitt Rivers Museum, University of Oxford; Professor Chris Scarre, Durham University; and Dr Sarah Shaw, Wolfson College, University of Oxford. Needless to say, whatever errors remain in the text are down to me.

Quotations from George Orwell in Chapter 9 are taken from 'Shooting an Elephant', in George Orwell, *Some Thoughts on the Common Toad* (London: Penguin Books, 2010). Those from Henry David Thoreau in Chapter 21 are taken from Henry David Thoreau, *Walden* (Oxford: Oxford

University Press, 1999). The poem by Lord Gorell in Chapter 25 is reproduced with the permission of his granddaughter and literary executor, Henrietta Gill. For Chapter 26, quotations by Robert Graves are taken from Robert Graves, *Goodbye to All That* (London: Penguin Books, 2000).

This book, and the radio series that accompanies it, is just one part of a much larger project on which I've been working: *Media and the Making of the Modern Mind*. That larger project is funded by the Leverhulme Trust, which generously awarded me a two-year research fellowship, and I would like to record my thanks to the Trust for having the flexibility and openmindedness to allow this 'side project' to grow beyond its original scope and take on a life of its own. I hope that, when the larger project is also complete, the Trust will feel that this brief diversion has been worthwhile. I would also like to thank my colleagues in the School of Media, Film, and Music at the University of Sussex for allowing me to complete the research on both *Noise* and *Media and the Making of the Modern Mind* before taking up my new duties.

I said that without the radio series there would have been no book, but there would also have been no book had it not been for my marvellous agent, Caroline Dawnay of United Agents, who saw the project's potential and brought it swiftly to the attention of Daniel Crewe at Profile Books. It was a lucky connection, since Profile has been a supremely efficient and professional publisher in all respects from beginning to end. Similarly, I must thank Joy Harris of the Joy Harris Literary Agency in New York for bringing *Noise* to the attention of Hilary Redmon at Ecco Books. Ecco, too, have been wonderful publishers. So I would like to thank, as well as Joy and Hilary, others at Ecco and HarperCollins who have been involved in producing this US edition, especially Christopher Smith, Shanna Milkey, Emma Janaskie, and Ashley Garland.

In the way of these things, it has once again been my immediate family – Henrietta, Eloise and Morgan – who have borne the heaviest burden of all during the past year or so. Writing a thirty-episode radio series and a thirty-chapter book at one and the same time has been, to say the least, all-consuming. Henrietta, Eloise and Morgan have helped me cope with the stresses and strains. They have covered for my absences and forgiven me for my distracted state. It's true, as always, that I couldn't have done it without them. And for that reason this book is dedicated to them with love.

Notes

Introduction

1. The phrase is from the British physicist G. W. C. Kaye, quoted in Karin Bijsterveld, *Mechanical Sound: Technology, Culture, and Public Problems of Noise in the Twentieth Century* (Cambridge, Mass., & London: MIT Press, 2008), p. 240.
2. Quoted in Emily Thompson, *The Soundscape of Modernity: Architectural Acoustics and the Culture of Listening in America, 1900–1933* (Cambridge, Mass., & London: MIT Press, 2004), p. 132.
3. John Cage, 'The Future of Music: Credo' (1937), in Richard Kostelanetz (ed.), *John Cage: An Anthology* (New York: De Capo Press, 1991).
4. Quoted in Thompson, *The Soundscape of Modernity*, pp. 143–4.
5. Elizabeth Foyster, 'Sensory Experiences: Smells, Sounds, and Touch', in Elizabeth Foyster and Christopher A. Whatley (eds), *A History of Everyday Life in Scotland* (Edinburgh: Edinburgh University Press, 2010), p. 217.
6. Hillel Schwarz, *Making Noise: From Babel to the Big Bang and*

Beyond (New York: Zone Books, 2011); Veit Erlmann, *Reason and Resonance: A History of Modern Aurality* (New York: Zone Books, 2010); Mike Goldsmith, *Discord: The Story of Noise* (Oxford: Oxford University Press, 2012).

7.   Thompson, *The Soundscape of Modernity*, p. 1.

8.   Dan MacKenzie, *The City of Din: A Tirade against Noise* (London: Adlard & Son, Bartholomew Press, 1916), pp. 1, 25.

9.   R. Murray Schafer, *The Soundscape: Our Sonic Environment and the Tuning of the World* (Rochester, Vt.: Destiny Books, 1994), p. 84.

10. Thompson, *The Soundscape of Modernity*, p. 2.

11. Douglas Kahn, *Noise, Water, Meat: A History of Sound in the Arts* (Cambridge, Mass., & London: MIT Press, 1999), p. 5.

1. Echoes in the Dark

1.   Ian Cross and Aaron Watson, 'Acoustics and the Human Experience of Socially-organized Sound', in Chris Scarre and Graeme Lawson (eds), *Archaeoacoustics* (Cambridge: McDonald Institute Monographs, 2006), pp. 107–16.

2.   *Ibid.*, p. 108–09.

3.   Iégor Reznikoff, 'Prehistoric Paintings, Sounds and Rocks', in Ellen Hickmann, Anne D. Kilmer, Ricardo Eichmann (eds), *Studien zur Musikarchäologie III* (Rahden: Verlag Marie Leidorf, 2002), pp. 44, 47–8.

4.   *Ibid.*, pp. 42–4, 47–8.

5.   Iégor Reznikoff, 'The Evidence of the Use of Sound Resonance from Palaeolithic to Medieval Times', in Scarre and Lawson, *Archaeoacoustics*, pp. 78–9.

6.   *Ibid.*, p. 80.

7.   Steven J. Waller, 'Intentionality of Rock-art Placement Deduced from Acoustical Measurements and Echo Myths', in Scarre and Lawson, *Archaeoacoustics*, pp. 31–9.

8.   Ian Cross, 'Lithoacoustics – Music in Stone: Preliminary Report', http://www.mus.cam.ac.uk/~ic108/lithoacoustics/ (October 2000).

9.   Cross and Watson, 'Acoustics and the Human Experience', in Scarre and Lawson, *Archaeoacoustics*, p. 113.

10. *Ibid.*, p. 114.

11. Steven J. Waller, 'Psychoacoustic influences of the echoing

environments of prehistoric art', Paper for the Acoustical Society of America, Cancun (November 2002).

12. Waller, 'Intentionality of Rock-art Placement', in Scarre and Lawson, *Archaeoacoustics*, p. 35.

13. David Lewis-Williams, *The Mind in the Cave: Consciousness and the Origins of Art* (London: Thames and Hudson, 2002), pp. 148–9.

14. Chris Scarre, 'Sound, Place and Space: Towards an Archaeology of Acoustics', in Scarre and Lawson, *Archaeoacoustics*, pp. 1–10.

15. Cross, 'Lithoacoustics'.

16. Francesco D'Errico and Graeme Lawson, 'The Sound Paradox: How to Assess the Acoustic Significance of Archaeological Evidence?', in Scarre and Lawson, *Archaeoacoustics*, pp. 41–58.

17. Ezra B. W. Zubrow and Elizabeth C. Blake, 'The Origin of Music and Rhythm', in Scarre and Lawson, *Archaeoacoustics*, pp. 117–126.

2. The Beat of Drums

1. James Gleick, *The Information: A History, a Theory, a Flood* (London: Fourth Estate, 2011), pp. 15, 18.

2. See, for instance, references to 'Drums made of a piece of a hollow Tree, covered on one end with any green Skin, and stretched with Thouls or Pins', which had been used 'in their Wars at home in Africa', in Hans Sloane, *A Voyage to the Islands Madera, Barbados, Nieves, S. Christophers and Jamaica* (London: 1701), vol. 1, p. liii. See also Richard Cullen Rath, *How Early America Sounded* (Ithaca, N.Y., and London: Cornell University Press, 2003), p. 78.

3. Roger T. Clarke, 'The Drum Language of the Tumba People', *American Journal of Sociology*, 40:1 (July 1934), p. 39.

4. *Ibid.*, p. 40.

5. Gleick, *The Information*, pp. 22–7.

6. See Steven Mithen, *The Singing Neanderthals: The Origins of Music, Language, Mind, and Body* (London: Phoenix, 2006), Chapter 6, 'Talking and Singing to Baby'.

7. Mithen, *The Singing Neanderthals*, Chapter 2, 'More than Cheesecake?'

8. Martin Clayton, Rebecca Sager and Udo Will, 'In Time with the Music: The Concept of Entrainment and Its Significance for Ethnomusicology', *European Meetings in Ethnomusicality*, 11 (2005).

9. Ian Cross, 'Music and Cognitive Evolution', in Louise Barrett and Robin Dunbar (eds), *Handbook of Evolutionary Psychology* (Oxford: Oxford University Press, 2007), pp. 26–8.

3. The Singing Wilderness

1. Sigurd F. Olson, *Listening Point* (1958; reprint, Minneapolis, Minn.: University of Minnesota Press, 2001), pp. 7–8. See also Peter Coates, 'The Strange Stillness of the Past: Toward an Environmental History of Sound and Noise', *Environmental History*, 10:4 (October 2005).

2. Bernie Krause, *The Great Animal Orchestra* (London: Profile Books, 2012), p. 51; R. Murray Schafer, *The Soundscape: Our Sonic Environment and the Tuning of the World* (Rochester, Vt.: Destiny Books, 1994), p. 23.

3. Quoted in Schafer, *The Soundscape*, pp. 22–3.

4. John Muir, *The Mountains of California* (1894), in *The Eight Wilderness Discovery Books* (London: Diadem Books, 1992), p. 399.

5. Krause, *The Great Animal Orchestra*, pp. 46–7.

6. Colin Turnbull, *The Forest People* (London: Picador, 1976), pp. 17–18.

7. Krause, *The Great Animal Orchestra*, p. 10.

8. Schafer, *The Soundscape*, pp. 18–19.

9. Krause, *The Great Animal Orchestra*, pp. 41–3.

10. *Ibid.*, pp. 27–30.

11. Steven Feld, *Sound and Sentiment: Birds, Weeping, Poetics, and Song as Kaluli Expression* (Philadelphia, Penn.: University of Pennsylvania Press, 1982), pp. 62, 144–50.

12. Feld, *Sound and Sentiment*. See also Steven Feld, 'Doing Anthropology in Sound', *American Ethnologist*, 31:4 (2004).

13. Marina Roseman, 'Healing Sounds from the Malaysian Rainforest' (1991), referenced in David Howes (ed.), *Sensual Relations: Engaging the Senses in Culture and Social Theory* (Ann Arbor, Mich.: University of Michigan Press, 2003), pp. 38–40.

14. Steven Mithen, *The Singing Neanderthals: The Origins of Music, Language, Mind, and Body* (London: Phoenix, 2006), Chapter 11, 'Imitating Nature'.

15. Claudette Kemper Columbus, 'Soundscapes in Andean Contexts', *History of Religions*, 44:2 (November 2004).

16. Stacie King and Gonzalo Sanchez Santiago, 'Soundscapes of the Everyday in Ancient Oaxaca, Mexico', *Archaeologies: Journal of the World Archaeological Congress* (2011).

17. S. Houston and K. Taube, 'An Archaeology of the Senses: Perception and Cultural Expression in Ancient Mesoamerica', *Cambridge Archaeological Journal*, 10 (2000), pp. 261–94.

4. A Ritual Soundscape

1. Aaron Watson, 'The Sounds of Transformation: Acoustics, Monuments and Ritual in the British Neolithic', in Neil S. Price (ed.), *The Archaeology of Shamanism* (London & New York: Routledge, 2001), pp. 178–9.

2. Aaron Watson, '(Un)intentional Sound? Acoustics and Neolithic Monuments', in Chris Scarre and Graeme Lawson (eds), *Archaeoacoustics* (Cambridge: McDonald Institute Monographs, 2006), p. 12.

3. *Ibid.*, p. 13.

4. Anna Ritchie, *Prehistoric Orkney* (London: B. T. Batsford/Historic Scotland, 1995), pp. 55–8.

5. Watson, 'The Sounds of Transformation', in Price, *The Archaeology of Shamanism*, pp. 180–81.

6. Aaron Watson and David Keating, 'Architecture and Sound: An Acoustic Analysis of Megalithic Monuments in Prehistoric Britain', *Antiquity*, 73 (1999), p. 199.

7. Watson, 'The Sounds of Transformation', in Price, *The Archaeology of Shamanism*, pp. 181–2.

8. Paul Devereux, 'Ears and Years: Aspects of Acoustics and Intentionality in Antiquity', in Scarre and Lawson, *Archaeoacoustics*, p. 29.

9. Watson, 'The Sounds of Transformation', in Price, *The Archaeology of Shamanism*, pp. 187–8.

10. Francesco Benozzo, 'Sounds of the Silent Cave: An Ethnophilological Perspective on Prehistoric "Incubation"', http://www.continuitas.org/texts/benozzo_sounds.pdf (accessed 4 September 2012).

11. Benozzo, 'Sounds of the Silent Cave'.

12. Benozzo, 'Sounds of the Silent Cave'.

5. The Rise of the Shamans

1. David Lewis-Williams, *The Mind in the Cave: Consciousness and the Origins of Art* (London: Thames & Hudson, 2002), pp. 131–5.

2. Waldemar Bogoras, *The Jessup North Pacific Expedition, Volume VII: The Chukchee, Part II: Religion* (Leiden: Brill, 1907), p. 433.

3. *Ibid.*, p. 439.

4. *Ibid.*, p. 429.

5. *Ibid.*, pp. 277–81, 289–90.

6. David Lewis-Williams and David Pearce, *Inside the Neolithic Mind:*

Consciousness, Cosmos and the Realm of the Gods (London: Thames & Hudson, 2005), pp. 286–7; emphasis added.

7.   Bogoras, *The Jessup North Pacific Expedition*, p. 374; emphasis added.
8.   *Ibid.*, pp. 382–3.
9.   Lewis-Williams, *The Mind in the Cave*, p. 131. See also Lewis-Williams and Pearce, *Inside the Neolithic Mind*, pp. 8–9.
10. Carolyn Marino Malone, *Façade as Spectacle: Ritual and Ideology at Wells Cathedral* (Leiden & Boston: Brill, 2004). See also, for details in this chapter on Wells, Graeme Lawson, 'Large Scale–Small Scale: Medieval Stone Buildings, Early Medieval Timber Halls and the Problem of the Lyre', in Chris Scarre and Graeme Lawson (eds), *Archaeoacoustics* (Cambridge: McDonald Institute Monographs, 2006), and Jerry Sampson, *Wells Cathedral West Front: Construction, Sculpture and Conservation* (Stroud: Sutton Publishing, 1998).
11. R. Murray Schafer, *The Soundscape: Our Sonic Environment and the Tuning of the World* (Rochester, Vt.: Destiny Books, 1994), pp. 10–11.
12. *Ibid.*
13. Lewis-Williams and Pearce, *Inside the Neolithic Mind*, pp. 86–7, 288.

6. Epic Tales

1.   Rosalind Thomas, *Literacy and Orality in Ancient Greece* (Cambridge: Cambridge University Press, 1992).
2.   Edward Luttwak, 'Homer Inc', *London Review of Books*, 23 February 2012.
3.   Marshall T. Poe, *A History of Communications: Media and Society from the Evolution of Speech to the Internet* (Cambridge: Cambridge University Press, 2011), pp. 67–72.
4.   Robert L. Fowler, 'Who Wrote the Iliad?', *Times Literary Supplement*, 14 March 2012.
5.   Fowler, 'Who Wrote the Iliad?' The quotation is part of Fowler's interpretation of M. L. West, *The Making of the Iliad* (Oxford: Oxford University Press, 2011).
6.   Ian Morley, 'Hunter-Gatherer Music and Its Implications for Identifying Intentionality in the Use of Acoustic Space', in Chris Scarre and Graeme Lawson (eds), *Archaeoacoustics* (Cambridge: McDonald Institute Monographs, 2006), pp. 100–102.
7.   Poe, *A History of Communications*, p. 55.
8.   Plato, *Phaedrus*, trans. R. Hackforth (Cambridge: Cambridge University Press, 1972), p. 157.

9. See, for example, Walter Ong, *Orality and Literacy: The Technologizing of the Word* (London: Routledge, 2002), pp. 8–9, 23, 27–9, 41–4.

10. See, for example, Poe, *A History of Communications*, pp. 26–99; and Asa Briggs and Peter Burke, A *Social History of the Media* (Cambridge: Polity Press, 2002).

11. C. Mackenzie Brown, 'Purāna as Scripture: From Sound to Image of the Holy Word in Hindu Tradition', *History of Religions*, 86:1 (August 1986), pp. 68–86.

12. Stamis L. Vassilantonopoulos and John N. Mourjopoulos, 'A Study of Ancient Greek and Roman Theatre Acoustics', *Acta Acustica United with Acustica*, 89 (2003), pp. 123–36.

13. *Ibid.*, pp. 123–33. See also K. Chourmouziadou and J. Kang, 'Acoustic Evolution of Ancient Greek and Roman Theatres', *Applied Acoustics*, 69 (2008).

14. Peter D. Arnott, *Public and Performance in the Greek Theatre* (London and New York: Routledge, 1989), pp. 5–6.

15. *Ibid.*, pp. 6–11, 23, 75.

7. Persuasion

1. Sam Leith, *You Talkin' to Me? Rhetoric from Aristotle to Obama* (London: Profile Books, 2011), pp. 6–7.

2. Charlotte Higgins, 'The New Cicero', *Guardian*, 26 November 2008.

3. Leith, *You Talkin' to Me?*, p. 219.

4. Amanda Claridge, *Rome: An Oxford Archaeological Guide* (Oxford: Oxford University Press, 2010), pp. 75–7, 85–7.

5. Leith, *You Talkin' to Me?*, p. 6.

6. Higgins, 'The New Cicero'.

7. Leith, *You Talkin' to Me?*, p. 30.

8. Zadie Smith, 'Speaking in Tongues', *New York Review of Books*, 26 February 2009.

9. Peter D. Arnott, *Public and Performance in the Greek Theatre* (London & New York: Routledge, 1989), pp. 79–80.

10. Arnott, *Public and Performance in the Greek Theatre*, p. 81.

11. *Ibid.*

12. Quoted in Leith, *You Talkin' to Me?*, p. 175.

13. Leith, *You Talkin' to Me?*, pp. 173–4.

14. Cicero, *in Catilinam*, Speech 1, 1.16–1.17, Latin Texts and

Translations, Perseus under PhiloLogic website: http://perseus. uchicago.edu/perseus-cgi/citequery3.pl?dbname=PerseusLatinTexts &getid=1&query=Cic.%20Catil.%201.17

15. From Plato's *Republic*, quoted in Arnott, *Public and Performance in the Greek Theatre*, p. 82.

16. Charles Hirschkind, 'Hearing Modernity: Egypt, Islam, and the Pious Ear', in Veit Erlmann (ed.), *Hearing Cultures: Essays on Sound, Listening and Modernity* (Oxford: Berg, 2004), p. 137.

17. *Ibid.*, pp. 133–4.

18. *Ibid.*, p. 134.

19. Personal communications with Jowita Kramer of the Oriental Institute, University of Oxford, and Dr Sarah Shaw, Wolfson College, Oxford, 30 April 2012.

20. Marshall T. Poe, *A History of Communications: Media and Society from the Evolution of Speech to the Internet* (Cambridge: Cambridge University Press, 2011), pp. 26–7.

21. Rosalind Thomas, *Literacy and Orality in Ancient Greece* (Cambridge: Cambridge University Press, 1992), pp. 20–21.

22. Simon Goldhill, 'Introduction: Why Don't Christians Do Dialogue?', in Simon Goldhill (ed.), *The End of Dialogue in Antiquity* (Cambridge: Cambridge University Press, 2009), p. 2.

23. Charlotte Higgins, 'Who's the Modern Cicero?', 20 October 2009, *Guardian* blog, http://www.guardian.co.uk/culture/ charlottehigginsblog/2009/oct/20/classics-barack-obama

8. Babble: The Noisy, Everyday World of Ancient Rome

1. Garret Keizer, *The Unwanted Sound of Everything We Want: A Book about Noise* (New York: Public Affairs, 2010), p. 81.

2. Juvenal, *Satire III*, 232–67, 'And Then There's the Traffic'; *Satire III*, 268–314, 'And the Violence'. Translated by A. S. Kline, Poetry in Translation (web archive): http://www.poetryintranslation.com/ PITBR/Latin/JuvenalSatires3.htm#_Toc281039208

3. Seneca, Epistle 56.1–2, quoted in Jerry Toner, *Popular Culture in Ancient Rome* (Cambridge: Polity Press, 2009), p. 130.

4. Florence Dupont, *Daily Life in Ancient Rome*, trans. Christopher Woodall (Oxford: Blackwell, 1992), pp. 47–9.

5. Toner, *Popular Culture in Ancient Rome*, pp. 140–42.

6. *Ibid.*, p. 142.

7. *Ibid.*

8. Dio Chrysostom,*Orations*, 20.9–10, Loeb translation with minor alterations, quoted in Toner, *Popular Culture in Ancient Rome*, p. 131.

9. Ray Laurence, *Roman Passions: A History of Pleasure in Imperial Rome* (London: Continuum, 2009), p. 9.

10. *Ibid.*, p. 5.

11. Keizer, *The Unwanted Sound of Everything We Want*, p. 89.

12. Amanda Claridge, *Rome: An Oxford Archaeological Guide* (Oxford: Oxford University Press, 2010), p. 263.

13. Alex Marshall, *Beneath the Metropolis: The Secret Lives of Cities* (London: Constable, 2006), p. 104.

14. Toner, *Popular Culture in Ancient Rome*, p. 129.

15. Keizer, *The Unwanted Sound of Everything We Want*, pp. 89–90.

16. Toner, *Popular Culture in Ancient Rome*, p. 130.

17. *Ibid.*, pp. 138–9.

18. *Ibid.*, p. 129.

19. Laurence, *Roman Passions*, p. 34.

20. *Ibid.*, pp. 46–7; Toner, *Popular Culture in Ancient Rome*, pp. 146–52.

21. Quoted in Dupont, *Daily Life in Ancient Rome*, pp. 50–51.

9. The Roaring Crowd

1. Keith Hopkins and Mary Beard, *The Colosseum* (London: Profile, 2011), pp. 24–5, 41.

2. Amanda Claridge, *Rome: An Oxford Archaeological Guide* (Oxford: Oxford University Press, 2010), pp. 312–15.

3. Hopkins and Beard, *The Colosseum*, p. 100.

4. Alison Futrell, *The Roman Games: A Sourcebook* (Oxford: Blackwell, 2006), p. ix.

5. Harry Sidebottom, *Ancient Warfare: A Very Short Introduction* (Oxford: Oxford University Press, 2004), pp. 115–18, 31–4.

6. *Ibid.*, pp. 86–8.

7. Futrell, *The Roman Games*, p. 7.

8. Hopkins and Beard, *The Colosseum*, pp. 42, 94–5.

9. Sarva Daman Singh, *Ancient Indian Warfare with Special Reference to the Vedic Period* (Leiden: E. J. Brill, 1965), pp. 79–83; Alfred S. Bradford, *With Arrow, Sword, and Spear: A History of Warfare in the Ancient World* (Westport, Conn. & London: Praeger, 2001), pp. 125–8.

10. Futrell, *The Roman Games*, p. 8.

11. Hopkins and Beard, *The Colosseum*, pp. 31–4.

12. Ovid, *The Art of Love*, 1.135–70, quoted in Futrell, *The Roman Games*, p. 105.

13. Futrell, *The Roman Games*, p. 101.

14. *Ibid.*, p. 27; Hopkins and Beard, *The Colosseum*, p. 100. The descriptions from Pliny are in Books 7–8 of his *Natural History*. A detailed description of the whole show is in Mary Beard, *The Roman Triumph* (Cambridge, Mass.: Harvard University Press, 2009), pp. 15–31.

15. See Futrell, *The Roman Games*, p. 27.

16. Bradford, *With Arrow, Sword, and Spear*, p. 128.

17. George Orwell, 'Shooting an Elephant', in George Orwell, *Some Thoughts on the Common Toad* (London: Penguin, 2010), p. 96.

18. Futrell, *The Roman Games*, p. 27.

19. *Ibid.*

20. *Ibid.*, p. 25.

21. *Ibid.*, pp. 20, 37–8.

22. Hopkins and Beard, *The Colosseum*, p. 41.

10. The Ecstatic Underground

1. Judith Toms, 'Catacombs', in Amanda Claridge, *Rome: An Oxford Archaeological Guide* (Oxford: Oxford University Press, 2010), p. 454.

2. See also Peregrine Horden, 'Introduction', in Peregrine Horden (ed.), *Music as Medicine: The History of Music Therapy Since Antiquity* (Aldershot: Ashgate, 2000), pp. 4–7.

3. Claridge, *Rome*, pp. 320–23.

4. See Roger Beck, 'Ritual, Myth, Doctrine, and Initiation in the Mysteries of Mithras: New Evidence from a Cult Vessel', *Journal of Roman Studies*, 90 (2000), pp. 145–80.

5. Roger Stalley, *Early Medieval Architecture* (Oxford: Oxford University Press, 1999), p. 20.

6. James W. Ermatinger, *Daily Life of Christians in Ancient Rome* (Westport, Conn. & London: Greenwood Press, 2007), p. ix.

7. Jerry Toner, *Popular Culture in Ancient Rome* (Cambridge: Polity Press, 2009), pp. 160–61.

8. R. Murray Schafer, *The Soundscape: Our Sonic Environment and the Tuning of the World* (Rochester, Vt.: Destiny Books, 1994), pp. 25–6.

9. *Ibid.*, pp. 142–3.

10. Ermatinger, *Daily Life*, pp. 72, 147–9; Toner, *Popular Culture in Ancient Rome*, p. 125.

11. C. Mackenzie Brown, 'Purāna as Scripture: From Sound to Image of the Holy Word in Hindu Tradition', *History of Religions*, 86:1 (August 1986), pp. 68–86.

12. Percival Price, *Bells and Man* (Oxford: Oxford University Press, 1983).

13. *Ibid.*, pp. 3–4.

14. C. M. Woolgar, *The Senses in Late Medieval England* (New Haven, Conn., & London: Yale University Press, 2006), p. 70.

15. Barbara Ehrenreich, *Dancing in the Streets: A History of Collective Joy* (London: Granta Books, 2007), pp. 75–6.

16. Ermatinger, *Daily Life*, p. 167.

17. Stalley, *Early Medieval Architecture*, p. 25.

18. Ehrenreich, *Dancing in the Streets*, p. 65.

19. Quoted in Ehrenreich, *Dancing in the Streets*, p. 73.

20. *Ibid.*

21. Ermatinger, *Daily Life*, p. 148.

22. Ehrenreich, *Dancing in the Streets*, p. 74.

23. See Simon Goldhill, 'Introduction: Why Don't Christians Do Dialogue?', in Simon Goldhill (ed.), *The End of Dialogue in Antiquity* (Cambridge: Cambridge University Press, 2009).

11. The Bells

1. See R. Murray Schafer, *The Soundscape: Our Sonic Environment and the Tuning of the World* (Rochester, Vt.: Destiny Books, 1994), pp. 54–5, and Alain Corbin, *Village Bells: Sound and Meaning in the Nineteenth-Century French Countryside* (London: Macmillan, 1999), p. 5.

2. Percival Price, *Bells and Man* (Oxford: Oxford University Press, 1983), pp. 82–3.

3. *Ibid.*, pp. 83–106.

4. The details of monastic bell use are drawn from C. M. Woolgar, *The Senses in Late Medieval England* (New Haven, Conn., & London: Yale University Press, 2006), pp. 70ff.; and Price, *Bells and Man*, pp. 86–90, 106.

5. Price, *Bells and Man*, p. 118.

6. *Ibid.*, p. 85.

7. *Ibid.*, p. 117.

8. Quoted in *ibid.*, p. 13.

9. Woolgar, *The Senses in Late Medieval England*, p. 71.

10. Schafer, *The Soundscape*, p. 54.

11. Price, *Bells and Man*, p. 119.
12. *Ibid.*, p. 128.
13. *Ibid.*, p. 15.
14. *Ibid.*, p. 122.
15. *Ibid.*, pp. 83–4.
16. Quoted in *ibid.*, p. 124.
17. Woolgar, *The Senses in Late Medieval England*, p. 72; Price, *Bells and Man*, p. 112.
18. Price, *Bells and Man*, p. 114.

12. Tuning the Body

1. *Gesta Romanorum, or, Entertaining Moral Stories, translated from the Latin with Preliminary Observations and Copious Notes by the Revd Charles Swan*, Vol. II (London: C. & J. Rivington, 1824), p. 34.
2. *Ibid.*
3. *Ibid.*, p. 35.
4. C. M. Woolgar, *The Senses in Late Medieval England* (New Haven, Conn., & London: Yale University Press, 2006), pp. 11–16; Christopher Page, 'Music and Medicine in the Thirteenth Century', in Peregrine Horden (ed.), *Music as Medicine: The History of Music Therapy Since Antiquity* (Aldershot: Ashgate, 2000), pp. 109–19.
5. Page, 'Music and Medicine', in Horden, *Music as Medicine*, pp. 116–17.
6. Woolgar, *The Senses in Late Medieval England*, pp. 75–7.
7. *Ibid.*, p. 87.
8. *Ibid.*, p. 77.
9. *Ibid.*, p. 88.
10. Rosemary Horrox (ed.), *The Black Death* (Manchester & New York: Manchester University Press, 1994), pp. 150–53.
11. Robert of Avesbury, from the chronicle 'De gestis mirabilibus regis Edwardi Tertii', quoted in W. O. Hassall (ed.), *Medieval England as Viewed by Contemporaries* (New York: Torch Books, 1965), pp. 157–8.
12. Woolgar, *The Senses in Late Medieval England*, p. 88. See also Norman Cohn, *The Pursuit of the Millennium*, revised edn (London: Secker & Warburg, 1970).
13. Robert Bartlett, *The Natural and the Supernatural in the Middle Ages* (Cambridge: Cambridge University Press, 2008).
14. Jamie James, *The Music of the Spheres: Music, Science and the Natural Order of the Universe* (London: Abacus, 1995), pp. 30–31, 59,

69–78; Woolgar, *The Senses in Late Medieval England*, p. 23; Charles
Burnett, 'Sound and Its Perception in the Middle Ages', in Charles
Burnett, Michael Fend and Penelope Gouk (eds), *The Second Sense:
Studies in Hearing and Musical Judgement from Antiquity to the
Seventeenth Century* (London: The Warburg Institute, 1991), pp. 43–
69; Peregrine Horden, 'Musical Solutions: Past and Present in Music
Therapy', in Horden, *Music as Medicine*.

15. Page, 'Music and Medicine', in Horden, *Music as Medicine*, pp. 110–
11.
16. Horden, 'Musical Solutions', in Horden, *Music as Medicine*, pp. 23–4.
17. *Ibid.*, pp. 9–10.

13. Heavenly Sounds

1. C. M. Woolgar, *The Senses in Late Medieval England* (New Haven,
Conn., & London: Yale University Press, 2006), p. 66.
2. This was experienced by me directly in September 2012, during
a day spent recording there for the BBC Radio 4 series. I was
accompanied by Matt Thompson, Professor Iégor Reznikoff and the
singer Dominique LeConte, and during our visit we recorded both
the artificially amplified sounds of the nuns of Vézelay singing and
the sound of Reznikoff and LeConte singing in ancient scales as they
walked around the nave. The 'harmonic' effect – totally free of echo
and yet highly resonant – was indeed stunning.
3. Iégor Reznikoff, 'The Evidence of the Use of Sound Resonance
from Palaeolithic to Medieval Times', in Chris Scarre and Graeme
Lawson (eds), *Archaeoacoustics* (Cambridge: McDonald Institute
Monographs, 2006), pp. 77–83.
4. *Ibid.*, p. 82.
5. Roger Stalley, *Early Medieval Architecture* (Oxford: Oxford
University Press, 1999), pp. 191–2; Reznikoff, 'The Evidence of the
Use of Sound Resonance', in Scarre and Lawson, *Archaeoacoustics*,
pp. 81–3.
6. Stalley, *Early Medieval Architecture*, pp. 116–19.
7. Interview with author, September 2012.
8. Woolgar, *The Senses in Late Medieval England*, p. 66; Thomas Forrest
Kelly, *Early Music: A Very Short Introduction* (Oxford: Oxford
University Press, 2011), pp. 20–24.
9. Paul Calamia and Jonas Braasch, 'Musical Granite Pillars in Ancient
Hindu Temples', *Journal of the Acoustical Society of America*, 123:5
(2008), p. 3604.

10. Peter Robb, *A History of India*, 2nd edn (Basingstoke: Palgrave Macmillan, 2011), pp. 75–7.

11. These are described in detail in Deborah Howard and Laura Moretti, *Sound and Space in Renaissance Venice: Architecture, Music, Acoustics* (New Haven, Conn., & London: Yale University Press, 2009). The various details in this and the following two paragraphs are drawn from this fascinating book, especially pp. 5, 17–21 and 27–42.

12. *Ibid.*, pp. 27–8.

13. *Ibid.*, p. 39.

14. *Ibid.*, pp. 5, 20–23.

15. Ebru Boyar and Kate Fleet, *A Social History of Istanbul* (Cambridge: Cambridge University Press, 2010), pp. 47–64.

16. Robb, *A History of India*, pp. 70, 77.

14. Carnival

1. Jonathan Sterne, 'Quebec's #Casseroles: On Participation, Percussion and Protest', *Sounding Out!* blog, 4 June 2012, http://soundstudiesblog.com/2012/06/04/casseroles/ (accessed 22 October 2012). See also Jonathan Sterne and Natalie Zemon Davis, 'Quebec's Manifs Casseroles are a Call for Order', *Globe and Mail*, 31 May 2012 (accessed 22 October 2012).

2. http://www.globalnoise.net (accessed 22 October 2012).

3. Quoted in Barbara Ehrenreich, *Dancing in the Streets: A History of Collective Joy* (London: Granta Books, 2007), p. 83.

4. *Ibid.*, pp. 78–80.

5. Mikhail Bakhtin, *Rabelais and His World* (New York: Wiley, 1984), p. 75.

6. E. P. Thompson, *Customs in Common* (London: Penguin Books, 1993), p. 51.

7. Bruce R. Smith, *The Acoustic World of Early Modern England: Attending to the O-Factor* (Chicago & London: University of Chicago Press, 1999), p. 133.

8. *Ibid.*, pp. 133–4.

9. *Ibid.*, p. 164.

10. *Ibid.*, pp. 154–5.

11. Sterne, 'Quebec's #Casseroles'.

12. Smith, *The Acoustic World of Early Modern England*, pp. 144–5.

13. *Ibid.*, p. 134.

14. Martin Ingram, 'Ridings, Rough Music and the "Reform of Popular Culture" in Early Modern England', *Past and Present*, 105 (1984),

p. 82; David Underdown, *Rebel, Riot, and Rebellion: Popular Politics and Culture in England, 1603–1660* (Oxford: Oxford University Press, 1985), p. 58. See also Emmanuel Le Roy Ladurie, *Carnival in Romans: Mayhem and Massacre in a French City* (London: Phoenix, 2003).

15. Thompson, *Customs in Common*, p. 68.
16. Leonardo Cardoso, 'Sound Politics in Sao Paulo, Brazil', *Sounding Out!* blog, 15 October 2012, http://soundstudiesblog. com/2012/10/15/sound-politics-in-sao-paulo-brazil/ (accessed 22 October 2012). See also Andrea Medrado, 'The Waves of the Hills: Community and Radio in the Everyday Life of a Brazilian Favela', unpublished PhD thesis, University of Westminster, 2010.

15. Restraint

1. Hannah Woolley, *The Gentlewoman's Companion, Or, a Guide to the Female Sex: The Complete Text of 1675, with an Introduction by Caterina Albano* (London: Prospect Books, 2001), pp. 109–14.
2. *Ibid.*, p. 79.
3. *Ibid.*, pp. 78–9.
4. *Ibid.*, p. 80.
5. *Ibid.*
6. *Ibid.*, p. 77.
7. Roger Thompson, *Sex in Middlesex: Popular Mores in a Massachusetts County, 1649–1699* (Amherst, Mass.: University of Massachusetts Press, 1989), pp. 157–65.
8. Woolley, *The Gentlewoman's Companion*, p. 69.
9. See Barbara Ehrenreich, *Dancing in the Streets: A History of Collective Joy* (London: Granta Books, 2007), pp. 97–100.
10. Keith Thomas, 'The Place of Laughter in Tudor and Stuart England', *Times Literary Supplement*, 21 January 1977, pp. 77–81.
11. Quoted in Thomas, 'The Place of Laughter', p. 80.
12. Woolley, *The Gentlewoman's Companion*, pp. 90–99.
13. Emily Cockayne, *Hubbub: Filth, Noise & Stench in England 1600–1770* (New Haven, Conn., & London: Yale University Press, 2007), p. 121.
14. *Ibid.*, pp. 11, 110–111.
15. *Ibid.*, pp. 115–16.
16. *Ibid.*, p. 115.
17. Woolley, *The Gentlewoman's Companion*, p. 72.
18. *Ibid.*, p. 77.

19. Richard Cullen Rath, *How Early America Sounded* (Ithaca, N.Y., & London: Cornell University Press, 2003), pp. 19–20.
20. *Ibid.*, p. 129.
21. Ehrenreich, *Dancing in the Streets*, pp. 119–20.
22. Woolley, *The Gentlewoman's Companion*, pp. 115–16.
23. *Ibid.*, p. 96.
24. Robert Darnton, *The Great Cat Massacre and Other Episodes in French Cultural History* (New York: Basic Books, 1984), p. 133.
25. E. P. Thompson, *Customs in Common* (London: Penguin Books, 1993), pp. 56–7.

16. Colonists

1. William Strachey, *A True Reportory of the Wracke and Redemption of Sir Thomas Gates, Knight; upon and from the Ilands of the Bermudas: his coming to Virginia, and the estate of that Colonie then, and after, under the government of the Lord La Warre, July 15, 1610, Volumes I– IV* (London: William Stansby, 1625).
2. *Ibid.*
3. *Ibid.*
4. *Ibid.*
5. Richard Cullen Rath, *How Early America Sounded* (Ithaca, N.Y., & London: Cornell University Press, 2003), pp. 51–3.
6. Strachey, *A True Reportory*.
7. R. Murray Schafer, *The Soundscape: Our Sonic Environment and the Tuning of the World* (Rochester, Vt.: Destiny Books, 1994), p. 47.
8. *Ibid.*, p. 47.
9. Rath, *How Early America Sounded*, p. 55.
10. *Ibid.*, pp. 55–7, 60–68.
11. Mary Rowlandson, *A Narrative of the Captivity, Sufferings, and Removes of Mrs. Mary Rowlandson* (Boston, Mass.: Massachusetts Sabbath School Society, 1856), pp. 9–11, 20, 24.
12. Rath, *How Early America Sounded*, p. 148.
13. *Ibid.*, pp. 7–8, 156–7.
14. *Ibid.*, p. 19.
15. Quoted in *ibid.*, p. 59.
16. *Ibid.*, pp. 148–9.
17. John Oxley, *Journals of Two Expeditions into the Interior of New South Wales, by Order of the British Government in the Years 1817–18* (Web edition: http://ebooks.adelaide.edu.au/o/oxley/john/o95j/), p. 22.

18. *Ibid.*, p. 29. See also Diane Collins, 'Acoustic Journeys: Exploration and the Search for an Aural History of Australia', *Australian Historical Studies*, 37:128 (2006), pp. 1–17.

19. Collins, 'Acoustic Journeys', p. 8.

20. *Ibid.*, pp. 11–14.

21. *Ibid.*, p. 11.

22. *Ibid.*

23. Major T. L. Mitchell, *Three Expeditions into the Interior of Eastern Australia with Description of the Recently Explored Region of Australia Felix and of the Present Colony of New South Wales, Vol. I* (London: T. & W. Boone, 1839), p. 109.

24. Collins, 'Acoustic Journeys', p. 6.

17. Shutting In

1. *Edinburgh Life in the Eighteenth Century, with an Account of the Fashions and Amusements of Society, Selected and Arranged from 'Captain Topham's Letters'* (Edinburgh: William Brown, 1899), pp. 2–3.

2. Jan-Andrew Henderson, *The Town Below Ground: Edinburgh's Legendary Underground City* (Edinburgh & London: Mainstream Publishing, 2008), p. 20.

3. Quoted in Henderson, *The Town Below Ground*, p. 28.

4. *Edinburgh Life in the Eighteenth Century*, pp. 2–3.

5. See Amanda Vickery, *Behind Closed Doors: At Home in Georgian England* (New Haven, Conn., & London: Yale University Press, 2009), p. 34.

6. *Edinburgh Life in the Eighteenth Century*, pp. 3–4.

7. Charles McKean, 'Improvement and Modernisation in Everyday Enlightenment Scotland', in Elizabeth Foyster and Christopher A. Whatley (eds), *A History of Everyday Life in Scotland* (Edinburgh: Edinburgh University Press, 2010), pp. 52–4.

8. Henderson, *The Town Below Ground*, p. 30; Emily Cockayne, *Hubbub: Filth, Noise & Stench in England 1600–1770* (New Haven, Conn., & London: Yale University Press, 2007), p. 122.

9. *Edinburgh Life in the Eighteenth Century*, p. 2.

10. *Ibid.*, pp. 19–20.

11. Elizabeth Foyster, 'Sensory Experiences: Smells, Sounds, Touch', in Foyster and Whatley, *A History of Everyday Life in Scotland*, pp. 217–33.

12. *Ibid.*, p. 226.

13. Henderson, *The Town Below Ground*, p. 29.
14. Vickery, *Behind Closed Doors*, p. 31.
15. *Ibid.*, pp. 26, 32–3.
16. *Edinburgh Life in the Eighteenth Century*, p. 11.
17. R. Murray Schafer, *The Soundscape: Our Sonic Environment and the Tuning of the World* (Rochester, Vt.: Destiny Books, 1994), p. 61.
18. John Macdonald, *Memoirs of an Eighteenth Century Footman: John Macdonald Travels 1745–1779* (London & New York: Routledge Curzon, 2005), p. 12.
19. *Ibid.*, p. 12.
20. Quoted in Henderson, *The Town Below Ground*, pp. 61–2.
21. Hugh M. Milne (ed.), *Boswell's Edinburgh Journals 1767–1786* (Edinburgh: Mercat Press, 2001), pp. 51, 265, 292.
22. Robert Darnton, *The Great Cat Massacre and Other Episodes in French Cultural History* (New York: Basic Books, 1984), pp. 75–104.
23. Cockayne, *Hubbub*, pp. 121, 129.
24. McKean, 'Improvement and Modernisation', pp. 63–4; Cockayne, *Hubbub*, p. 128.
25. McKean, 'Improvement and Modernisation', p. 64.
26. Leah Leneman, *Alienated Affections: The Scottish Experience of Divorce and Separation, 1684–1830* (Edinburgh: Edinburgh University Press, 1998), p. 4; McKean, 'Improvement and Modernisation', pp. 52, 65.
27. *Edinburgh Life in the Eighteenth Century*, p. 7.
28. McKean, 'Improvement and Modernisation', p. 67.
29. Cockayne, *Hubbub*, p. 128.
30. *Ibid.*, p. 130.

18. Master and Servant

1. Quoted in Leah Leneman, *Alienated Affections: The Scottish Experience of Divorce and Separation, 1684–1830* (Edinburgh: Edinburgh University Press, 1998), pp. 31–2.
2. Amanda Vickery, *Behind Closed Doors: At Home in Georgian England* (New Haven, Conn., & London: Yale University Press, 2009), pp. 7, 33–5.
3. Emily Cockayne, *Hubbub: Filth, Noise & Stench in England 1600–1770* (New Haven, Conn., & London: Yale University Press, 2007), p. 120.
4. *Ibid.*
5. *Ibid.*

6.   Vickery, *Behind Closed Doors*, pp. 14–18, 24–6.
7.   *Ibid.*, pp. 27, 38.
8.   John L. Locke, *Eavesdropping: An Intimate History* (Oxford: Oxford University Press, 2010), pp. 172–4.
9.   Vickery, *Behind Closed Doors*, p. 38.
10. Locke, *Eavesdropping*, pp. 83–4.
11. *Edinburgh Life in the Eighteenth Century, with an Account of the Fashions and Amusements of Society, Selected and Arranged from 'Captain Topham's Letters'* (Edinburgh: William Brown, 1899), pp. 20–21.
12. All details of this case and quotations are from *Trials for Adultery: Or, The History of Divorces, Being Select Trials at Doctors Commons, for Adultery, Cruelty, Fornication, Impotence, &c., From the Year 1760, to the Present Time, Including Whole of the Evidence of each Cause, Vol. VI* (London: 1780), pp. 1–22.
13. Leneman, *Alienated Affections*, pp. 31–4.
14. Locke, *Eavesdropping*, pp. 176–8.
15. *Ibid.*, p. 186.
16. *Ibid.*, pp. 113, 139–40, 175–9.
17. Vickery, *Behind Closed Doors*, p. 27; Locke, *Eavesdropping*, p. 187.
18. *Ibid.*, p. 187.
19. Quoted in *ibid.*, pp. 27–9.

19. Slavery and Rebellion

1.   Quoted in Mark M. Smith (ed.), *Stono: Documenting and Interpreting a Southern Slave Revolt* (Columbia, S.C.: University of Southern Carolina Press, 2005), p. 4.
2.   *Ibid.*, p. 7.
3.   *Ibid.*, p. 9.
4.   *Ibid.*, p. 12.
5.   Peter Charles Hoffer, *Cry Liberty: The Great Stono River Slave Rebellion of 1739* (Oxford: Oxford University Press, 2010).
6.   'Account of the Negroe Insurrection in South Carolina', unknown author, *c.* October 1739, reprinted in Smith, *Stono*, pp. 13–15.
7.   Hoffer, *Cry Liberty*, pp. 102–6.
8.   Hans Sloane, *A Voyage to the Islands Madera, Barbados, Nieves, S. Christophers and Jamaica, Vol. I* (London, 1701), pp. xlvii–xlviii.
9.   *Ibid.*, pp. xlvii–xlviii, lii.
10. *Ibid.*, pp. xlviii–xlix.
11. *Ibid.*, p. lii.

12. Richard Cullen Rath, *How Early America Sounded* (Ithaca, N.Y., & London: Cornell University Press, 2003), p. 79.
13. *Ibid.*, pp. 78–9.
14. *Ibid.*, p. 77.
15. John K. Thornton, 'African Dimensions of the Stono Rebellion', *American Historical Review*, 96 (1991), pp. 1103–13.
16. Hoffer, *Cry Liberty*, p. 157. See also John Hope Franklin and Loren Schweninger, *Runaway Slaves: Rebels on the Plantation* (Oxford & New York: Oxford University Press, 1999), pp. 11–12.
17. Rath, *How Early America Sounded*, p. 89.
18. *Ibid.*
19. *Ibid.*, p. 91.
20. *Ibid.*, pp. 91–3.
21. Shane White and Graham White, 'Listening to Southern Slavery', in Mark M. Smith (ed.), *Hearing History: A Reader* (Athens, Ga., & London: University of Georgia Press, 2004), pp. 247–66.
22. Frederick Law Olmsted, *A Journey in the Back Country* (New York: Mason Brothers, 1861), pp. 188–9.
23. Hoffer, *Cry Freedom*, p. 47.
24. Caryl Phillips, *The Atlantic Sound* (London: Vintage, 2001).
25. Steven Feld, *Jazz Cosmopolitanism in Accra: Five Musical Years in Ghana* (Durham, N.C., & London: Duke University Press, 2012), pp. 2–10.

20. Revolution and War

1. Quoted in Simon Schama, *Citizens: A Chronicle of the French Revolution* (London: Penguin, 1989), p. 238.
2. Quoted in Mark M. Smith, *Listening to Nineteenth-Century America* (Chapel Hill, N.C., & London: University of North Carolina Press, 2001), p. 201.
3. Schama, *Citizens*, pp. 114, 151.
4. *Ibid.*, p. 115.
5. Original French text in Pierre Constant, *Musique des fêtes et cérémonies de la révolution française* (Paris: Imprimerie Nationale, 1899), pp. 146–7. Translation from Roy Rosenzweig Center for History and New Media, George Mason University: http://chnm.gmu.edu/revolution/browse/songs/#
6. Quoted in Schama, *Citizens*, p. 316.
7. *Ibid.*, pp. 324–5.
8. *Ibid.*

9. *Ibid.*, pp. 138–9, 325.

10. *Ibid.*, p. 326.

11. See George Rudé, *The Crowd in History* (London: Serif, 1995), pp. 93–133.

12. Smith, *Listening to Nineteenth-Century America*, p. 198.

13. *Ibid.*, p. 202.

14. The Smithsonian Institution has a film clip of veterans performing their 'rebel yell': http://www.smithsonianmag.com/video/What-Did-the-Rebel-Yell-Sound-Like.html

15. Smith, *Listening to Nineteenth-Century America*, pp. 196, 201.

16. *Ibid.*, p. 205.

17. Charles D. Ross, 'Sight, Sound, and Tactics in the American Civil War', in Mark M. Smith (ed.), *Hearing History: A Reader* (Athens, Ga., & London: University of Georgia Press, 2004), p. 270.

18. *Ibid.*, pp. 275–6.

19. *Ibid.*

21. The Conquering Engines: Industrial Revolution

1. Henry David Thoreau, *Walden* (Oxford: Oxford University Press, 1999), p. 102.

2. *Ibid.*, pp. 116–17.

3. *Ibid.*, pp. 112–13.

4. *Ibid.*, p. 106.

5. R. Murray Schafer, *The Soundscape: Our Sonic Environment and the Tuning of the World* (Rochester, Vt.: Destiny Books, 1994), p. 81.

6. Thoreau, *Walden*, p. 109.

7. *Ibid.*, p. 108.

8. Jeremy Black, *A History of the British Isles*, 2nd edn (Basingstoke: Palgrave Macmillan, 2003), p. 201.

9. Charles Dickens, *Dombey and Son* (London: Bradbury and Evans, 1848), p. 219.

10. *Ibid.*, p. 233.

11. Schafer, *The Soundscape*, p. 73. See also, for an American perspective on the same changes, Leo Marx, *The Machine in the Garden: Technology and the Pastoral Ideal* (London & New York: Oxford University Press, 2000).

12. Hugh Miller, *First Impressions of England and Its People* (Edinburgh: Adam & Charles Black, 1861), pp. 156–7.

13. *Ibid.*

14. Letter of Thomas Carlyle to Alexander Carlyle, 11 August 1824, quoted in Humphrey Jennings, *Pandaemonium 1660–1886: The Coming of the Machine as Seen by Contemporary Observers*, ed. Mary-Lou Jennings and Charles Madge (London: André Deutsch, 1985), p. 165.

15. Quoted in Dan McKenzie, *The City of Din: A Tirade Against Noise* (London: Adlard & Son, Bartholomew Press, 1916), p. 90.

16. Quoted in Schafer, *The Soundscape*, p. 75.

17. McKenzie, *The City of Din*, p. 91.

18. Miller, *First Impressions*, pp. 157–8.

19. *Ibid.*

20. *Ibid.*, pp. 25–30.

21. Paddy Scannell and David Cardiff, *A Social History of British Broadcasting, Vol. 1 1922–1939: Serving the Nation* (Oxford: Basil Blackwell, 1991), p. 342.

22. *Ibid.*

23. Schafer, *The Soundscape*, p. 63. See also Lewis Mumford, *Technics and Civilization* (London: Routledge, 1934), p. 201.

24. Schafer, *The Soundscape*, pp. 75–7.

25. *Ibid.*, p. 79.

22. The Beat of a Heart, the Tramp of a Fly

1. Quoted in Edward Shorter, *Doctors and Their Patients* (New Brunswick & London: Transaction, 1991), p. 49.

2. J. Worth Estes, 'Dropsy', in Kenneth F. Kiple (ed.), *The Cambridge World History of Human Disease* (Cambridge: Cambridge University Press, 1993), Cambridge Histories Online (accessed 19 November 2012), DOI:10.1017/CHOL9780521332866.101.

3. Shorter, *Doctors and Their Patients*, pp. 49–50.

4. *Ibid.*, p. 84.

5. Quoted in Jonathan Sterne, 'Mediate Auscultation, the Stethoscope, and the "Autopsy of the Living": Medicine's Acoustic Culture', *Journal of Medical Humanities*, 22:2 (2001), p. 119.

6. *Ibid.*, p. 118.

7. *Ibid.*, p. 117.

8. Shorter, *Doctors and Their Patients*, p. 76.

9. Sterne, 'Mediate Auscultation', pp. 128–9.

10. Kate Flint, *The Victorians and the Visual Imagination* (Cambridge: Cambridge University Press, 2008), pp. 13–19.

11. Quoted in Malcolm Nicolson, 'Having the Doctor's Ear in Nineteenth-Century Edinburgh', in Mark M. Smith (ed.), *Hearing History: A Reader* (Athens, Ga., & London: University of Georgia Press, 2004), p. 157. See also W. F. Bynum and Roy Porter (eds), *Medicine and the Five Senses* (Cambridge: Cambridge University Press, 1993).

12. Nicolson, 'Having the Doctor's Ear', in Smith, *Hearing History*, p. 160.

13. *Ibid.*

14. *Ibid.*

15. *Ibid.*, p. 152; Shorter, *Doctors and Their Patients*, pp. 31–4.

16. *Ibid.*, p. 83.

17. John M. Picker, *Victorian Soundscapes* (Oxford & New York: Oxford University Press, 2003), p. 3.

18. Quoted in Picker, *Victorian Soundscapes*, p. 3.

19. *Ibid.*, p. 4.

20. *Ibid.*, p. 6.

23. The New Art of Listening

1. John Muir, *My First Summer in the Sierra* (London: Constable, 1911), pp. 255–6.

2. *Ibid.*, pp. 247–8.

3. John Muir, *Our National Parks* (1901), in *The Eight Wilderness Discovery Books* (London: Diadem Books, 1992), pp. 459, 544.

4. John Picker, *Victorian Soundscapes* (Oxford: Oxford University Press, 2003), p. 7.

5. *Ibid.*, p. 8.

6. Peter Gay, *The Bourgeois Experience, Victoria to Freud, Vol. IV: The Naked Heart* (New York & London: Norton, 1995), p. 24.

7. *Ibid.*, p. 14.

8. *Ibid.*

9. *Ibid.*, pp. 15–18.

10. *Ibid.*, p. 13.

11. *Ibid.*, p. 27.

12. *Ibid.*, p. 31.

13. See Peter Watson, *The German Genius: Europe's Third Renaissance, the Second Scientific Revolution, and the Twentieth Century* (London & New York: Simon & Schuster, 2010), especially pp. 65–88, 198.

14. Gay, *The Bourgeois Experience*, p. 19.

15. *Ibid.*, pp. 19–20.

16. *Ibid.*, pp. 20–21.
17. *Ibid.*, p. 21.
18. *Ibid.*, p. 35.
19. James H. Johnson, *Listening in Paris: A Cultural History* (Berkeley, Los Angeles & London: University of California Press, 1995), p. 233.
20. Gay, *The Bourgeois Experience*, p. 19.
21. Charles Lamb, 'A Chapter on Ears', in *The Essays of Elia* (London: Dent, 1923), p. 46. See also Picker, *Victorian Soundscapes*, p. 8.
22. Johnson, *Listening in Paris*, p. 236.
23. Kate Flint, *The Victorians and the Visual Imagination* (Cambridge: Cambridge University Press, 2000), pp. 64–78, 86.
24. Oliver Lodge, *Past Years: An Autobiography* (London: Hodder & Stoughton, 1931), pp. 174, 345–6.

24. Life in the City

1. Letter, Thomas Carlyle to Margaret A. Carlyle, 31 December 1852, *Collected Letters*, Vol. 27, pp. 387–8, The Carlyle Letters Online, http://carlyleletters.dukejournals.org
2. Letter, Jane Carlyle to Isabella Carlyle, 24 November 1841, *Collected Letters*, Vol. 13, pp. 307–8.
3. Letter, Jane Carlyle to Grace Welsh, 23 February 1842, *Collected Letters*, Vol. 14, pp. 49–50.
4. Letter, Thomas Carlyle to Margaret A. Carlyle, 12 March 1853, *Collected Letters*, Vol. 28, pp. 73–4; letter, Thomas Carlyle to Jane Carlyle, 8 July 1853, *Collected Letters*, Vol. 28, pp. 185–7.
5. Letter, Thomas Carlyle to Margaret A. Carlyle, 11 July 1853, *Collected Letters*, Vol. 28, pp. 196–8.
6. *Ibid.*
7. Letter, Jane Carlyle to Kate Sterling, 19 November 1853, *Collected Letters*, Vol. 28, pp. 318–19; letter, Jane Carlyle to Charles Redwood, 25 December 1853, *Collected Letters*, Vol. 28, pp. 350–51.
8. Garret Keizer, *The Unwanted Sound of Everything We Want: A Book about Noise* (New York: Public Affairs, 2010), p. 109.
9. John Picker, *Victorian Soundscapes* (Oxford: Oxford University Press, 2003), p. 42.
10. *Ibid.*, pp. 46–7.
11. Edwin Hopewell-Ash, *On Keeping Our Nerves in Order* (London: Mills & Boon, 1928).
12. Dan McKenzie, *The City of Din: A Tirade against Noise* (London: Adlard & Son, Bartholomew Press, 1916), pp. 32–3, 105–8.

13. *Ibid.*, pp. 32–3.
14. *Ibid.*, pp. 33–4.
15. *Ibid.*, pp. 38, 63.
16. *Ibid.*, p. 52.
17. James Ford, *Slums and Housing, with Special Reference to New York City: History, Conditions, Policy* (Cambridge, Mass.: Harvard University Press, 1936), pp. 526–7.
18. Katherine Greider, *The Archaeology of Home: An Epic Set on a Thousand Square Feet of the Lower East Side* (New York: Public Affairs, 2011), p. 143.
19. Jacob A. Riis, *How the Other Half Lives* (London: Penguin Books, 1997), p. 91.
20. Jacob A. Riis, *Children of the Tenements* (New York & London: Macmillan, 1903), pp. 35, 127–9, 208–9.
21. Lawrence J. Epstein, *At the Edge of a Dream: The Story of Jewish Immigrants on New York's Lower East Side 1880–1920* (New York: Wiley, 2007), p. 45.
22. Riis, *How the Other Half Lives.*
23. Greider, *The Archaeology of Home*, p. 144.
24. Riis, *Children of the Tenements*, p. 135.
25. *Ibid.*, pp. 3–4.
26. Riis, *How the Other Half Lives*, p. 10.
27. *Ibid.*, p. 11.
28. Linda Granfield and Arlene Alda, *97 Orchard Street, New York: Stories of Immigrant Life* (New York: Tundra Books, 2001).
29. Riis, *How the Other Half Lives*, pp. 84–5.
30. *Ibid.*, p. 101.
31. *New York Herald*, 13 May 1894, quoted in Greider, *The Archaeology of Home*, p. 165.
32. Greider, *The Archaeology of Home*, pp. 164–9.
33. Riis, *How the Other Half Lives*, p. 14.
34. *Ibid.*, pp. 122–3, 128–31.

25. Capturing Sound

1. 'Voice Mail Delivers, Retains Final Words', *St Petersburg Times*, 8 September 2002.
2. The Sonic Memorial Project: http://www.sonicmemorial.org/sonic/public/about.html
3. Michael Chanan, *Repeated Takes: A Short History of Recording and Its Effects on Music* (London & New York: Verso, 1995), p. 1.

4. William Howland Kenney, *Recorded Music in American Life: The Phonograph and Popular Memory, 1890–1945* (Oxford & New York: Oxford University Press, 1999), pp. 23–4.

5. Carolyn Marvin, *When Old Technologies Were New: Thinking about Electric Communication in the Late Nineteenth Century* (Oxford & New York: Oxford University Press, 1988), pp. 73–4.

6. Quoted in Jacques Attali, *Noise: The Political Economy of Music* (Manchester: Manchester University Press, 1985), p. 91.

7. Quoted in Susan Douglas, *Listening In: Radio and the American Imagination, from Amos 'n' Andy and Edward R. Murrow to Wolfman Jack and Howard Stern* (New York: Times Books, 1999), p. 46.

8. Evan Eisenberg, *The Recording Angel: Explorations in Phonography* (New York: McGraw-Hill, 1987), p. 57.

9. John Picker, *Victorian Soundscapes* (Oxford: Oxford University Press, 2003), p. 133.

10. Ronald Gorell, 'To a Gramophone in an African Camp', 28 August 1910, in Lord Gorell, *1904–1936 Poems* (London: John Murray, 1937), p. 20.

11. Quoted in Steven Connor, 'Megaphonics', BBC Radio 3, 28 February 1997.

12. *The Times*, 16 June 1914.

13. Chanan, *Repeated Takes*, pp. 10–12.

14. Chanan, *Repeated Takes*, pp. 7, 26–30, 37, 48–50; Kenney, *Recorded Music in American Life*, pp. 40–42.

15. Chanan, *Repeated Takes*, pp. 15–16.

16. *Ibid.*, p. 15.

17. Halifu Osumare, *The Hiplife in Ghana: West African Indigenization of Hip-Hop* (New York: Palgrave Macmillan, 2012), p. 1.

18. Kenney, *Recorded Music in American Life*, pp. 18–19.

19. Chanan, *Repeated Takes*, p. 15.

20. Quoted in Chanan, *Repeated Takes*, p. 51.

21. *Ibid.*

22. *Ibid.*, pp. 19, 52–3.

26. Shell Shock

1. Ian Passingham, *Pillars of Fire: The Battle of Messines Ridge, June 1917* (London: The History Press, 2012), p. 102.

2. Robert Graves, *Goodbye to All That* (London: Penguin Books, 2000), p. 92.

3. Paul Fussell, *The Great War and Modern Memory* (Oxford: Oxford University Press, 2000), pp. 36–74; Yaron Jean, 'The Sonic Mindedness of the Great War: Viewing History through Auditory Lenses', in Florence Feiereisen and Alexandra Merley Hill (eds), *Germany in the Loud Twentieth Century* (Oxford: Oxford University Press, 2012), pp. 51–62.

4. Erich Maria Remarque, *All Quiet on the Western Front* (New York: Random House, 1958), p. 106.

5. *Ibid.*, p. 59; Graves, *Goodbye to All That*, pp. 82–3.

6. Quoted in Peter Watson, *German Genius: Europe's Third Renaissance, the Second Scientific Revolution, and the Twentieth Century* (London & New York: Simon & Schuster, 2010), p. 550.

7. Graves, *Goodbye to All That*, p. 127.

8. *Ibid.*, p. 83.

9. Jean, 'The Sonic Mindedness of the Great War', in Feiereisen and Hill, *Germany in the Loud Twentieth Century*, p. 54.

10. *Ibid.*

11. *Ibid.*, p. 51; Graves, *Goodbye to All That*, p. 83.

12. Graves, *Goodbye to All That*, p. 102.

13. *Ibid.*, p. 96. See also R. Murray Schafer, *The Soundscape: Our Sonic Environment and the Tuning of the World* (Rochester, Vt.: Destiny Books, 1994), pp. 8–9.

14. Graves, *Goodbye to All That*, p. 176.

15. *Ibid.*, pp. 142–3; Jean, 'The Sonic Mindedness of the Great War', in Feiereisen and Hill, *Germany in the Loud Twentieth Century*, p. 54.

16. Graves, *Goodbye to All That*, p. 98.

17. Anton Schnack, 'In a Shellhole', quoted in Watson, *German Genius*, p. 551.

18. Graves, *Goodbye to All That*, p. 133.

19. *Ibid.*

20. Schafer, *The Soundscape*, p. 9.

21. Peter Barham, *Forgotten Lunatics of the Great War* (New Haven, Conn., & London: Yale University Press, 2007), p. 16.

22. Edwin L. Ash, *Nerve in Wartime* (London: Mills & Boon, 1914), pp. 24–6.

23. Charles Myers, *Shell Shock in France 1914–18* (Cambridge: Cambridge University Press, 1940), pp. 24–5.

24. *Ibid.*, p. 26.

25. Graves, *Goodbye to All That*, p. 143.

26. Barham, *Forgotten Lunatics*, p. 4.

27. *Ibid.*, p. 53.
28. *Ibid.*, p. 84.
29. Ash, *Nerve in Wartime*, p. 26.
30. Fiona Reid, *Broken Men: Shell Shock, Treatment and Recovery in Britain 1914–30* (London & New York: Continuum, 2010), p. 75.
31. Barham, *Forgotten Lunatics*, p. 43.
32. Edwin Lancelot Ash, *The Problem of Nervous Breakdown* (London: Mills & Boon, 1919), p. 220.
33. Barham, *Forgotten Lunatics*, pp. 18–21.
34. *Ibid.*, pp. 45–50.
35. See Jay Winter, 'Thinking about Silence', in Efrat Ben Ze'ev, Ruth Ginio and Jay Winter (eds), *Shadows of War: A Social History of Silence in the Twentieth Century* (Cambridge: Cambridge University Press, 2010), pp. 3–31.
36. Edwin L. Ash, *Nerves and the Nervous*, revised edn (London: Mills & Boon, 1921), p. 21.
37. Graves, *Goodbye to All That*, p. 235. See also Michael Roper, *The Secret Battle: Emotional Survival in the Great War* (Manchester: Manchester University Press, 2009).
38. Jean, 'The Sonic Mindedness of the Great War', in Feiereisen and Hill, *Germany in the Loud Twentieth Century*, p. 60.
39. Ash, *Nerves and the Nervous*, p. 11.

27. Radio Everywhere!

1. 'Aerial Voices', *London Standard*, 28 December 1912.
2. David Hendy, 'The Dreadful World of Edwardian Wireless', in Siân Nicholas and Tom O'Malley (eds), *Moral Panics, Social Fears, and the Media: Historical Perspectives* (London & New York: Routledge, 2013), pp. 76–89.
3. Richard Evans, *The Third Reich in Power* (London: Penguin Books, 2006), p. 121.
4. Corey Ross, *Media and the Making of Modern Germany: Mass Communications, Society, and Politics from the Empire to the Third Reich* (Oxford: Oxford University Press, 2008), p. 330.
5. Ralf Georg Reuth, *Goebbels*, trans. Krishna Winston (London: Constable, 1993), pp. 176–7.
6. *Ibid.*, p. 176; Evans, *The Third Reich*, p. 135; Ross, *Media and the Making of Modern Germany*, p. 330.
7. *Ibid.*, pp. 330–31.
8. Leon Trotsky, 'Radio, Science, Technique and Society', *Labour*

Review, 2:6 (November–December 1957). The speech was originally delivered by Trotsky in Moscow on 1 March 1926.

9. William Hard, quoted in Michele Hilmes, 'British Quality, American Chaos: Historical Dualisms and What They Leave Out', *Radio Journal: International Studies in Broadcast and Audio Media*, 1:1 (2003), p. 13.

10. Michele Hilmes, *Radio Voices: American Broadcasting 1922–1952* (Minneapolis & London: University of Minnesota Press, 1997); Susan Douglas, *Listening In: Radio and the American Imagination* (New York: Times Books, 1999).

11. Gerd Horten, *Radio Goes to War: The Cultural Politics of Propaganda during World War II* (Berkeley & Los Angeles: University of California Press, 2003), p. 17; David Ryfe, 'From Media Audience to Media Public: A Study of Letters Written in Reaction to FDR's Fireside Chats', *Media Culture & Society*, 23 (2001), pp. 767–81.

12. J. C. W. Reith, *Broadcast over Britain* (London: Hodder & Stoughton, 1924), pp. 34, 64.

13. *Ibid.*, pp. 161–2. See also David Hendy, 'BBC Radio Four and Conflicts over Spoken English in the 1970s', *Media History*, 12:3 (2006), pp. 273–89.

14. Hadley Cantril and Gordon W. Allport, *The Psychology of Radio* (New York & London: Harper & Brothers Publishers, 1935), p. 20.

15. *Ibid.*, pp. 109–25.

16. *Ibid.*, pp. 259–60.

17. Ryfe, 'From Media Audience to Media Public', p. 770; Horten, *Radio Goes to War*, pp. 51–2.

18. Greg Goodale, *Sonic Persuasion: Reading Sound in the Recorded Age* (Chicago: University of Illinois Press, 2011), p. xi.

19. Quoted in Horten, *Radio Goes to War*, p. 52.

20. Quoted in Ross, *Media and the Making of Modern Germany*, p. 331.

21. *Ibid.*, pp. 334–40.

22. Paddy Scannell and David Cardiff, *A Social History of British Broadcasting, Volume 1, 1922–1939: Serving the Nation* (Oxford: Basil Blackwell, 1991), p. 375.

28. Music While You Shop, Music While You Work

1. Quoted in Joseph Lanza, *Elevator Music: A Surreal History of Muzak, Easy-Listening and other Moodsong* (London: Quartet Books, 1995), p. 17.

2. *Ibid.*, p. 18.
3. Antony Copley, *Music and the Spiritual: Composers and Politics in the 20th Century* (London: Ziggurat Books, 2012).
4. See Jonathan Sterne, 'Sounds Like the Mall of America: Programmed Music and the Architectonics of Commercial Space', *Ethnomusicology*, 41 (1997), pp. 22–50.
5. Lanza, *Elevator Music*, p. 39.
6. *Ibid.*, pp. 21, 27–8.
7. Quoted in *ibid.*, pp. 36–7.
8. On Fritz Lang's *Metropolis*, see Anton Kaes, 'Metropolis (1927): City, Cinema, Modernity', in Noah Isenberg (ed.), *Weimar Cinema* (New York: Columbia University Press, 2009), pp. 173–91; Ian Roberts, *German Expressionist Cinema: The World of Light and Shadow* (London & New York: Wallflower, 2008). On the relationship between the machine aesthetics of cinema and sound in the interwar years see James Mansell, 'Rhythm, Modernity and the Politics of Sound', in Scott Anthony and James Mansell (eds), *The Projection of Britain: A History of the GPO Film Unit* (Basingstoke: Palgrave Macmillan, 2011), pp. 161–7.
9. Lanza, *Elevator Music*, p. 27.
10. Christina L. Baade, *Victory through Harmony: The BBC and Popular Music in World War II* (Oxford: Oxford University Press, 2012).
11. *Ibid.*, pp. 65–6.
12. *Manchester Evening Chronicle*, 9 March 1943, quoted in Baade, *Victory through Harmony*, p. 60.
13. *Ibid.*, pp. 62–7.
14. *Ibid.*, pp. 70–71.
15. *Ibid.*, pp. 63–79.
16. Lanza, *Elevator Music*, pp. 2–4.
17. Quoted in *ibid.*, pp. 28–9.
18. Quoted in *ibid.*, pp. 2–4.
19. George Prochnik, *In Pursuit of Silence: Listening for Meaning in a World of Noise* (New York: Anchor Books, 2010), pp. 111–12.

29. An Ever Noisier World

1. Emily Thompson, *The Soundscape of Modernity: Architectural Acoustics and the Culture of Listening in America, 1900–1933* (Cambridge, Mass., & London: MIT Press, 2004), p. 148.
2. 'Noise', *Saturday Review of Literature*, 1, quoted in Thompson, *The Soundscape of Modernity*, p. 117.

3. Garret Keizer, *The Unwanted Sound of Everything We Want: A Book about Noise* (New York: Public Affairs, 2010).

4. 'Noise', *Saturday Review of Literature*, 1, quoted in Thompson, *The Soundscape of Modernity*, pp. 119–20.

5. *Ibid.*, pp. 155–7.

6. *Ibid.*, pp. 118–27, 157–8.

7. *Ibid.*, pp. 124–5.

8. Andrea Medrado, 'The Waves of the Hills: Community and Radio in the Everyday Life of a Brazilian Favela', unpublished PhD thesis, University of Westminster, 2010.

9. See, for example, http://ama.ghanadistricts.gov. gh/?arrow=nws&read=45694 and http://www.ghananewsagency. org/details/Social/Dansoman-residents-call-on-AMA-to-act-against-noisy-church/?ci=4&ai=35356

10. Steven Connor, *Noise*, BBC Radio 3, 27 February, 1997.

11. Keizer, *The Unwanted Sound of Everything We Want*, pp. 7–8.

12. *Ibid.*, p. 131; Bernie Krause, *The Great Animal Orchestra* (London: Profile Books, 2012), pp. 188–93.

13. *Ibid.*, p. 187.

14. Keizer, *The Unwanted Sound of Everything We Want*, p. 30.

15. *Ibid.*, p. 124. See also James Gleick, *The Information: A History, a Theory, a Flood* (London: Fourth Estate, 2011); John Brockman (ed.), *Is the Internet Changing the Way You Think?* (New York & London: Harper, 2011); Nicholas Carr, *The Shallows: How the Internet is Changing the Way We Read, Think and Remember* (London: Atlantic, 2011); Susan Jacoby, *The Age of American Unreason* (London: Old Street Publishing, 2008); and Nate Silver, *The Signal and the Noise: The Art and Science of Prediction* (London: Allen Lane, 2012).

16. Steven Connor, *Noise*, BBC Radio 3, 28 February, 1997.

17. Quoted in Thompson, *The Soundscape of Modernity*, p. 131.

18. *Ibid.*, pp. 131–2.

30. The Search for Silence

1. Lafcadio Hearn, *Japan: An Attempt at Interpretation* (1905), quoted by the Ashmolean Museum display in Room 36.

2. 'Making a Personal Retreat', New Melleray Abbey: http://www. newmelleray.org/retreatbrochure.asp?display=sub (accessed 14 December 2012).

3. George Prochnik, *In Pursuit of Silence: Listening for Meaning in a World of Noise* (New York: Anchor Books, 2010), p. 27.

4. *Ibid*.

5. Garret Keizer, *The Unwanted Sound of Everything We Want: A Book about Noise* (New York: Public Affairs, 2010), p. 132.

6. Emily Thompson, *The Soundscape of Modernity: Architectural Acoustics and the Culture of Listening in America, 1900–1933* (Cambridge, Mass., & London: MIT Press, 2004), pp. 174–83.

7. *Ibid.*, pp. 198–205.

8. Stanford Corbett, 'An Office Building of the New Era', *Scientific American* (December 1929), quoted in Thompson, *The Soundscape of Modernity*, p. 202.

9. Keizer, *The Unwanted Sound of Everything We Want*, p. 45.

10. Prochnik, *In Pursuit of Silence*, p. 197.

11. Keizer, *The Unwanted Sound of Everything We Want*, pp. 132–7.

12. Prochnik, *In Pursuit of Silence*, pp. 117–19.

13. *Ibid.*, pp. 221–3.

14. Walter Murch, 'Touch of Silence', in Larry Sider, Diane Freeman and Jerry Sider (eds), *Soundscape: The School of Sound Lectures, 1998–2001* (London: Wallflower Press, 2003), pp. 83–102.

15. Kevin Maguire, 'BBC Cheers Up Lonely Staff with the Chit-chat Machine', *Guardian*, 14 October 1999: http://www.guardian. co.uk/media/1999/oct/14/bbc.uknews

16. Ravi Mehta, Rui Zhu and Amar Cheema, 'Is Noise Always Bad? Exploring the Effects of Ambient Noise on Creative Cognition', *Journal of Consumer Research*, 39:4 (2012), pp. 784–99. See also Leo Hickman, 'Want to Get Creative? Then Visit a Coffee Shop', *Guardian*, 24 June 2012: http://www.guardian.co.uk/theguardian/ shortcuts/2012/jun/24/get-creative-visit-coffee-shop

17. Keizer, *The Unwanted Sound of Everything We Want*, p. 242.

Epilogue

1. Mark M. Smith, *Listening to Nineteenth-Century America* (Chapel Hill & London: University of North Carolina Press, 2001), p. 262.

2. *Ibid.*, p. 264.

3. Garret Keizer, *The Unwanted Sound of Everything We Want: A Book about Noise* (New York: Public Affairs, 2010), p. 34.

4. Michael Bull and Les Back, 'Introduction: Into Sound', in Michael Bull and Les Back (eds), *The Auditory Culture Reader* (Oxford & New York: Berg, 2003), p. 9.

5. Harry Mount, 'The Queen's Earplugs Are Just the Lead We Require in the Battle on Noise', *Daily Telegraph*, 30 August 2012.

6. Bull and Back, 'Introduction: Into Sound', in Bull and Back, *The Auditory Culture Reader*, p. 9.

7. Owen Jones, *Chavs: The Demonization of the Working Class* (London: Verso, 2011), p. 8.

8. Smith, *Listening to Nineteenth-Century America*, p. 266.

9. Joanna Bourke, *Fear: A Cultural History* (London: Virago, 2005), p. 353.

Index